Siemens
NX 2019
for Novices
(Continuous Release)
Learn By Doing

CADSoft Technologies

Siemens NX 2019 for Novices (Continuous Release)
Learn By Doing

CADSoft Technologies

CADSoft Technologies

ISBN 978-1-942689-55-3

DEDICATION

To students, who are dedicated to learning new technologies

THANKS

To employees of CADSoft Technologies
for their cooperation

Training Programs offered by CADSoft Technologies

CADSoft Technologies provides quality training on various software packages including Computer Aided Design, Manufacturing and Engineering (CAD/CAM/CAE), Web and Programming Languages, animation, architecture, and GIS. The training is delivered at its training centers or online at any time, place, and pace to individuals, corporates as well as the students of colleges, universities, and CAD/CAM/CAE training centers. The main features of this program are:

Training in Classroom Settings and Online Training

A team of highly experienced instructors and qualified engineers at CADSoft Technologies conduct the classes in classroom environment at its centers. This team has authored several textbooks that are rated "one of the best" in their categories and are used in various colleges, universities, and training centers in North America, Europe, and in other parts of the world.

CADSoft Technologies with its cost effective and time saving initiative strives to deliver online training in the comfort of your home or work place, thereby relieving you from the hassles of traveling to training centers. CADSoft Technologies strives to be the best training institute in USA. CADSoft Technologies has developed career oriented programs by following state-of-the-art teaching and learning methodologies.

Courses Offered by CADSoft Technologies

CADSoft Technologies conducts classes on the following software packages:

CAD/CAM/CAE: *CATIA, Pro/ENGINEER Wildfire, Creo Parametric, Creo Direct, SOLIDWORKS, Autodesk Inventor, Solid Edge, NX, AutoCAD, AutoCAD LT, AutoCAD Plant 3D, Customizing AutoCAD, SolidCAM, NX CAM, NX Mold, Creo Mold, Alias Design, Alias Automotive, Ansys Fluent, and ANSYS*

Architecture and GIS: *Autodesk Revit Architecture, AutoCAD Civil 3D, Autodesk Revit Structure, AutoCAD Map 3D, Revit MEP, Navisworks, Primavera, and Bentley STAAD Pro, MS Project, MX Road, ArcGIS, and Raster Design*

Animation and VFX: *Autodesk 3ds Max, VRay, Autodesk Maya, Blackmagic Design Fusion Studio, Adobe Premiere, Adobe Photoshop, Adobe Indesign, Adobe Illustrator, CorelDraw, The Foundry NukeX, and MAXON CINEMA 4D*

Web and Programming: *C/C++, HTML5/CSS3, JavaScript, jQuery, Bootstrap, PHP, MySQL, Dreamweaver, VB.NET, Oracle, AJAX, and Java*

Personality Development: *Personality Development and Engineering Ethics/Soft Skills Course*

Table of Contents

Dedication iii
Preface ix

Chapter 1: Introduction to NX

Introduction to NX 1-2
System Requirements 1-4
Getting Started with NX 1-4
Important Terms and Definitions 1-6

Chapter 2: Drawing Sketches for Solid Models

Introduction 2-2
Starting NX 2-2
Sketching Tools 2-3
- Drawing Sketches Using the Profile Tool 2-3
- Drawing Individual Lines 2-3
- Drawing Arcs 2-4
- Drawing Circles 2-4
- Drawing Rectangles 2-4
- Filleting Sketched Entities 2-4

Exiting the Sketch Environment 2-4
Tutorial 1 2-5
Tutorial 2 2-9
Exercise 1 2-13
Exercise 2 2-14

Chapter 3: Adding Geometric and Dimensional Constraints to Sketches

Constraining Sketches 3-2
Concept of Constrained Sketches 3-2
- Under-Constrain 3-2
- Fully-Constrain 3-2
- Over-Constrain 3-2

Dimensioning Sketches 3-3
Geometric Constraints 3-4
- Applying Additional Constraints Individually 3-4
- Showing All Constraints in a Sketch 3-5

Tutorial 1 3-5
Tutorial 2 3-11
Exercise 1 3-15
Exercise 2 3-16

Chapter 4: Editing, Extruding, and Revolving Sketches

Editing Sketches 4-2
    Trimming Sketched Entities 4-2
    Extending Sketched Entities 4-3
Creating Base Features by Extruding 4-4
Creating Solid Revolved Bodies 4-4
Tutorial 1 4-6
Tutorial 2 4-9
Exercise 1 4-11
Exercise 2 4-11

Chapter 5: Working with Datum Planes, Coordinate Systems, and Datum Axes

Additional Sketching and Reference Planes 5-2
Types of Datum Planes 5-3
    Creating Three Fixed (Principle) Datum Planes 5-3
    Creating Relative Datum Planes 5-3
Creating Datum Coordinate Systems 5-4
Tutorial 1 5-7
Tutorial 2 5-12
Exercise 1 5-16
Exercise 2 5-17

Chapter 6: Advanced Modeling Tools-I

Advanced Modeling Tools 6-2
Creating Holes by Using the Hole Tool 6-2
Creating Grooves 6-3
Creating Ribs 6-3
Creating Chamfers 6-3
Creating an Edge Blend 6-4
Tutorial 1 6-5
Tutorial 2 6-12
Exercise 1 6-19
Exercise 2 6-20

Chapter 7: Advanced Modeling Tools-II

Advanced Modeling Tools 7-2
Pattern Feature Tool 7-2
    Creating a Linear Pattern 7-2
    Creating a Circular Pattern 7-3
Sweeping Sketches Along the Guide Curves 7-5
Creating Threads 7-6
    Creating Symbolic Threads 7-7
    Creating Detailed Threads 7-8
Creating Shell Features 7-9

Creating Drafts 7-10
Tutorial 1 7-11
Tutorial 2 7-16
Exercise 1 7-22
Exercise 2 7-23

Chapter 8: Assembly Modeling-I

The Assembly Environment 8-2
Invoking the Assembly Environment 8-2
Invoking the Assembly Environment Using the New Dialog Box 8-2
Invoking the Assembly Environment in the Current Part File 8-2
Types of Assembly Design Approaches 8-3
Creating Bottom-up Assemblies 8-3
Placing Components in the Assembly Environment 8-4
Tutorial 1 8-6
Exercise 1 8-20

Chapter 9: Assembly Modeling-II

The Top-down Assembly Design Approach 9-2
Creating Components Using the Top-down Assembly Design Approach 9-2
Creating Components in Assembly File 9-2
Tutorial 1 9-4
Exercise 1 9-11

Chapter 10: Surface Modeling

Introduction to Surface Modeling 10-2
Invoking the Shape Studio Environment 10-2
Creating an Extruded Surface 10-3
Creating a Revolved Surface 10-3
Creating a Combined Surface 10-4
Creating a Surface Using the Through Curves Tool 10-6
Adding Thickness to a Surface 10-8
Tutorial 1 10-9
Exercise 1 10-14
Index **I-1**

This page is intentionally left blank

Preface

Siemens NX 2019 (Continuous Release)

Siemens NX, a product of SIEMENS Corp., is one of the world's leading CAD/CAM/CAE packages. Being a solid modeling tool, it not only unites 3D parametric features with 2D tools, but also addresses every design-through-manufacturing process. Besides providing an insight into the design content, the package promotes collaboration between companies and provides them an edge over their competitors.

In addition to creating solid models and assemblies, the 2D drawing views can also be generated easily in the **Drafting** environment of NX. The drawing views that can be generated include orthographic, section, auxiliary, isometric, and detail views. The model dimensions and reference dimensions in the drawing views can also be generated. The bidirectionally associative nature of this software ensures that the modifications made in the model are reflected in the drawing views and vice-versa. In NX, you can create sketches directly in the Modeling environment.

The **Siemens NX 2019 for Novices (Continuous Release): Learn By Doing** textbook has been written with the intention of helping the readers effectively use the solid modeling tools in NX. After reading this textbook, you can create solid models, surface models, and assemblies. The mechanical engineering industry examples and tutorials used in this book ensure that the users can relate the knowledge of this book with the actual mechanical industry designs. The main features of this textbook are as follows:

- **Tutorial Approach**

  The author has adopted the tutorial point-of-view and the learn-by-doing approach throughout the textbook. This approach guides the users through the process of creating the models in the tutorials.

- **Real-World Projects as Tutorials**

  The author has used about 15 real-world mechanical engineering projects as tutorials in this book. This enables the readers to relate the tutorials to the models in the mechanical engineering industry. In addition, there are about 16 exercises that are also based on the real-world mechanical engineering projects.

- **Tips and Notes**

  The additional information related to various topics is provided to the users in the form of tips and notes.

- **Learning Objectives**

  The first page of every chapter summarizes the topics that are covered in that chapter.

- **Exercises**

  Exercises are given at the end of the chapter and they can be used by the instructors to assess the knowledge of the student. **You need to perform all the exercises as output of some of these exercises may have been used as a reference later in this book.**

Formatting Conventions Used in the Textbook

Please refer to the following list for the formatting conventions used in this textbook.

- Names of tools, buttons, options, groups, and toolbar, are written in boldface. Example: The **Extrude** tool, the **OK** button, the **Feature** group, and so on.

- Names of dialog boxes, drop-downs, drop-down lists, list boxes, areas, edit boxes, check boxes, and radio buttons are written in boldface. Example: The **Edge Blend** dialog box, the **Surface Drop-down** of the **Create** group, the **Tangent Edges** drop-down list of the **Shell** dialog box, the **Distance** edit box of the **Extrude** dialog box, the **Symmetric** check box in the **Pattern Feature** dialog box, the **Plain** radio button of the **Edit Background** dialog box, and so on.

- Values entered in edit boxes are written in boldface. Example: Enter **5** in the **Pitch Distance** edit box.

- Names and paths of the files saved are italicized. Example: *c03tut03.prt*, *C:\NX_12.0\c03*, and so on.

- The methods of invoking a tool/option from the **Menu**, **Ribbon** are enclosed in a shaded box.

Ribbon:	Home > Standard > New
Menu:	File > New

Naming Conventions Used in the Textbook

Button

The item in a dialog box that has a 3D shape like a button is termed as **Button**. For example, **OK** button, **Cancel** button, **Apply** button, and so on.

Dialog Box

In this textbook, different terms are used for referring to the components of a dialog box. Refer to Figure 1 for the terminology used.

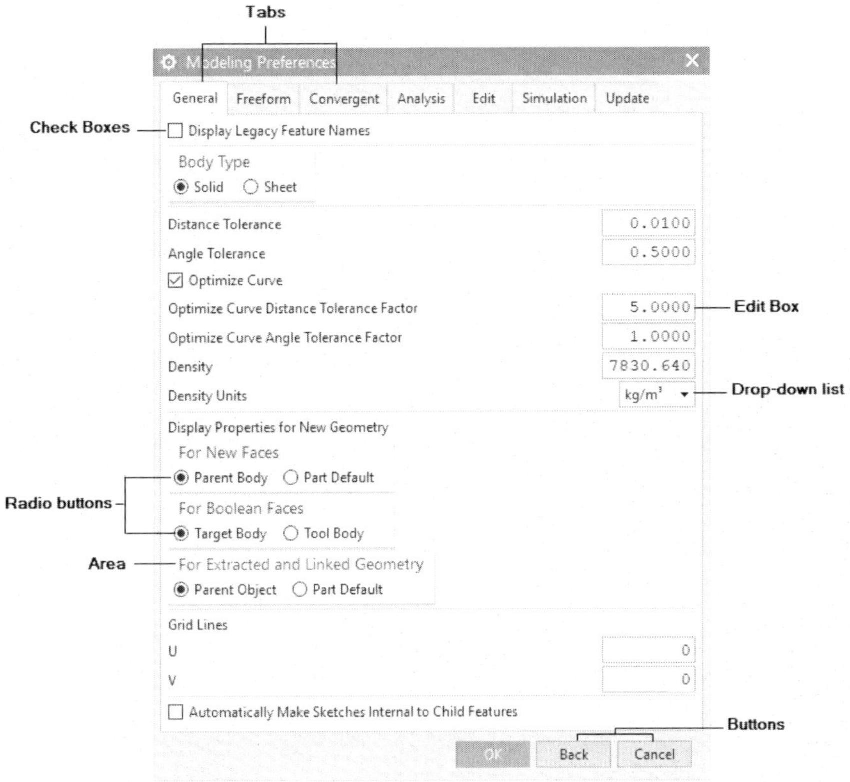

Figure 1 *The components in a dialog box*

Drop-down

A drop-down is the one in which a set of common tools are grouped together. You can identify a drop-down with a down arrow on it. These drop-downs are given a name based on the tools grouped in them. For example, the **Design Feature** drop-down, the **Blend** drop-down, and so on; refer to Figure 2.

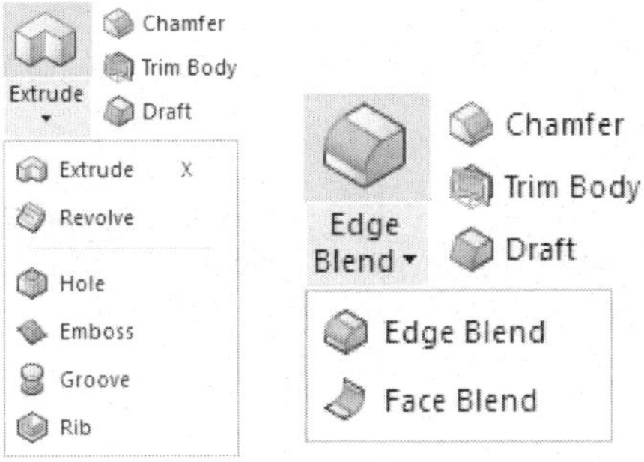

Figure 2 *The **Design Feature Drop-down** and **Blend Drop-down***

Gallery

A gallery is the one in which a set of common tools are grouped together. For example, **Detail Feature** gallery of the **More** gallery; refer to Figure 3.

***Figure 3** The **Detail Feature** gallery of the **More** gallery*

Drop-down List

A drop-down list is the one in which a set of options are grouped together. You can set various parameters using these options. You can identify a drop-down list with a down arrow on it. For example, **Boolean** drop-down list, **Layout** drop-down list, and so on; refer to Figure 4.

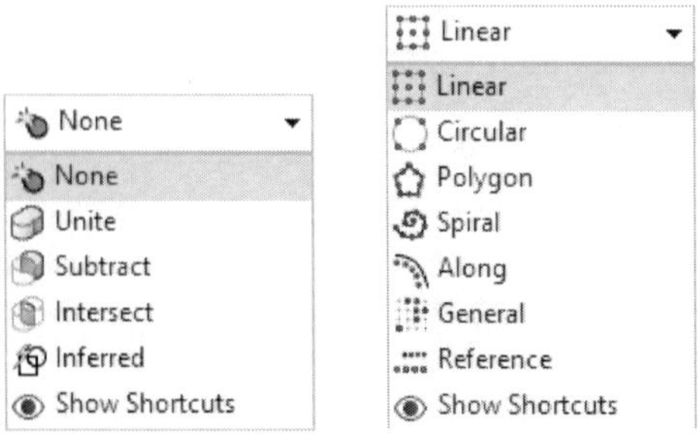

***Figure 4** The **Boolean** and **Layout** drop-down lists*

Options

Options are the items that are available in shortcut menu, drop-down list, dialog boxes, and so on. For example, choose the **Fit** option from the shortcut menu displayed on right-clicking in the drawing area; select the **Face** option from the drop-down list in the **Type** rollout; refer to Figure 5.

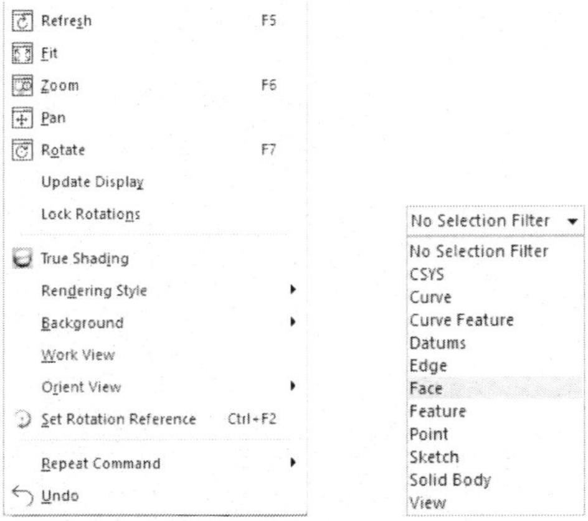

Figure 5 *Options in the shortcut menu and drop-down list in the* ***Type*** *rollout*

This page is intentionally left blank

Chapter 1

Introduction to NX

Learning Objectives

After completing this chapter, you will be able to:

- *Understand different environments in NX*
- *Understand the system requirements for NX*
- *Start a new file in NX*
- *Understand the important terms and definitions used in NX*

INTRODUCTION TO NX

Welcome to NX. As a new user of this software package, you will join hands with thousands of users of this high-end CAD/CAM/CAE/PLM tool. If already familiar with the previous releases, you can upgrade your designing skills with tremendous improvement in this latest release.

The latest release of NX introduces improved capabilities for convergent modeling, CAD design, drafting, tooling design, automation, design simulation, programming, data translation, validation, mold design, and much more. Active workspace in NX is built based on these items to provide seamless access to PLM capabilities right within NX.

NX, a product of SIEMENS Corp., is a completely re-engineered, next-generation family of CAD/CAM/CAE/PLM software solutions for Product Life Cycle Management. Through its exceptionally easy-to-use and state-of-the-art user interface, NX delivers innovative technologies for maximum productivity and creativity from the basic concept to the final product. NX reduces the learning curve by allowing flexibility in the use of feature-based and parametric designs.

The subject of interpretability offered by NX includes receiving legacy data from other CAD systems and even between its own product data management modules. The real benefit is that the links remain associative. As a result, any changes made to this external data are notified to you and the model can be updated quickly.

When you open an old file or start a new file in NX, you will enter the Gateway environment. It allows you to examine the geometry and drawing views that have been created. In the Gateway environment, you can invoke any environment of NX.

NX serves the basic design tasks by providing different environments. An environment is defined as a specific area, consisting of a set of tools which allows the user to perform specific design tasks in that particular area. You need to start the required environment after starting a new part file. As a result, you can invoke any environment of NX in the same working part file. The basic environments in NX are the Modeling environment, Shape Studio environment, Drafting environment, Assembly environment, and the Sheet Metal environment. These environments are discussed next.

Modeling Environment

The Modeling environment is a parametric and feature-based environment in which you can create solid models. The basic requirement for creating solid models in this environment is a sketch. You can draw the sketch directly in the Modeling environment by using the tools available in the **Direct Sketch** group of the **Home** tab. The sketch can also be drawn in the sketching environment. The sketching environment can be invoked by choosing the **Sketch** tool from the **Direct Sketch** group of the **Home** tab or by choosing the **Sketch** tool from the **Direct Sketch** group of the **Curve** tab. While drawing a sketch, various applicable constraints and dimensions are automatically applied to it. Additional constraints and dimensions can also be applied manually. After drawing the sketch, you need to convert the sketch into a feature using the tools available in the Modeling environment. You can create placed features such as fillets, chamfers, taper, and so on and can also assign materials to the models in this environment.

Shape Studio Environment

The Shape Studio environment is also a parametric and feature-based environment in which you can create surface models. The tools in this environment are similar to those in the Modeling environment. The only difference is that the tools in this environment are used to create basic and advanced surfaces. You are also provided with the surface editing tools which are used to manipulate the surfaces to obtain the required shape. This environment is useful for conceptual and industrial design.

Assembly Environment

The Assembly environment is used to assemble the components using the assembly constraints available in this environment. There are two types of assembly design approaches in NX, Bottom-up and Top-down.

In the bottom-up approach of the assembly, the previously created components are assembled together to maintain their design intent and in the top-down approach, components are created in the Assembly environment.

In the Assembly environment you can also assemble an existing assembly with the current assembly. The **Perform Analysis** tool provides the facility to check the interference and clearance between the components in an assembly.

Drafting Environment

The Drafting environment is used for the documentation of the parts or assemblies created earlier in the form of drawing views and their detailing. There are two types of drafting techniques: generative drafting and interactive drafting.

The generative drafting technique is used to automatically generate the drawing views of the parts and assemblies. The parametric dimensions added to the component in the Modeling environment during its creation can also be generated and displayed automatically in the drawing views. The generative drafting is bidirectionally associative in nature. If you modify the dimensions in the Drafting environment, the model will automatically update in the Modeling environment and vice-versa. You can also generate the Bill of Material (BOM) and balloons in the drawing views.

In interactive drafting, you need to create the drawing views by sketching them using the normal sketching tools and then adding the dimensions.

Sheet Metal Environment

The Sheet Metal application provides an environment for the design of sheet metal parts used in machinery, enclosures, brackets, and other parts normally manufactured with a brake press. The Sheet Metal application is intended mainly for designing parts with cylindrical bend regions but conical and curved bend regions are also possible. Generally, the sheet metal components are created to generate the flat pattern of a sheet, study the design of the dies and punches, study the process plan for designing, and the tools needed for manufacturing the sheet metal components.

SYSTEM REQUIREMENTS

System requirements that ensure the smooth running of NX are as follows:

- 64-bit - Windows 10 Operating System.
- 8GB of RAM is the minimum requirement but it is recommended to have 16GB or more RAM for all the applications to run smoothly.
- Java version - 1.8.0 or higher
- True Color (32-bit) or 16 million colors (24-bit)
- Screen Resolution: 1280 x 1024 or higher, widescreen format.

GETTING STARTED WITH NX

Install NX on your system and then start it by double-clicking on its shortcut icon on the desktop of your computer. After the system has loaded all the required files to start NX, the initial interface will be displayed, as shown in Figure 1-1.

Note

When you install NX on your system, its shortcut icon is not created by default on your desktop. You need to drag it out from the Start menu of the Windows.

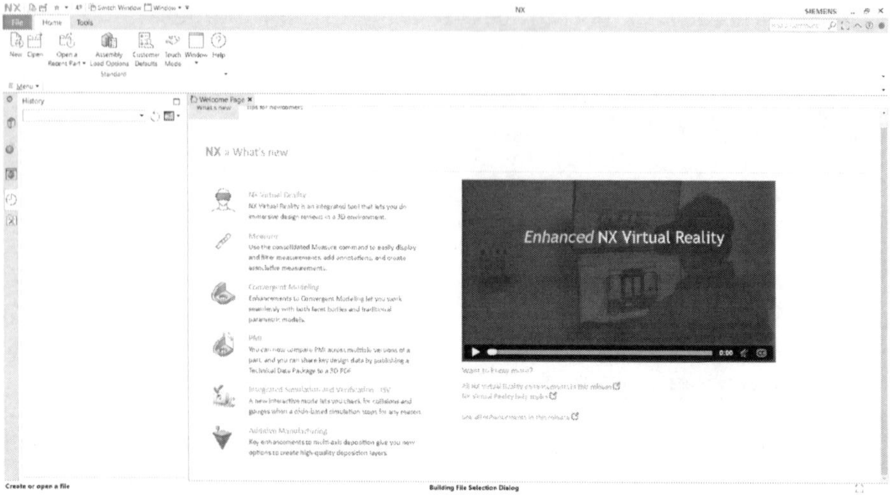

***Figure 1-1** The initial interface that appears after starting NX*

Tip

*1. In this release, by default, the user interface theme is set to **Light**. But, you can change the interface theme as per your requirement. To do so, choose **Menu > Preferences > User Interface** from the **Top Border Bar**; the **User Interface Preferences** dialog box will be displayed. Next, select the required option from the **Type** drop-down list available in the **NX Theme** group of the **Theme** node and choose the **OK** button.*

*2. To change the **Default Presentation of Dialog Content** of any dialog box, choose **Menu > Preferences > User interface** from the **Top Border Bar**; the **User Interface Preferences** dialog box will be displayed. Next, select the **Options** node available at the left of the dialog box; the **Dialog Boxes** group will be displayed. Now, select the **More** radio button in the **Default Presentation of Dialog Content** area and choose the **OK** button.*

Choose **File > New** from the **Ribbon**; the **New** dialog box will be displayed, as shown in Figure 1-2. Make sure that **Model** template is selected in the **Templates** rollout of the dialog box. Next, enter the name of the file in the **Name** edit box and choose the **OK** button; the Modeling environment will be displayed, refer to Figure 1-3.

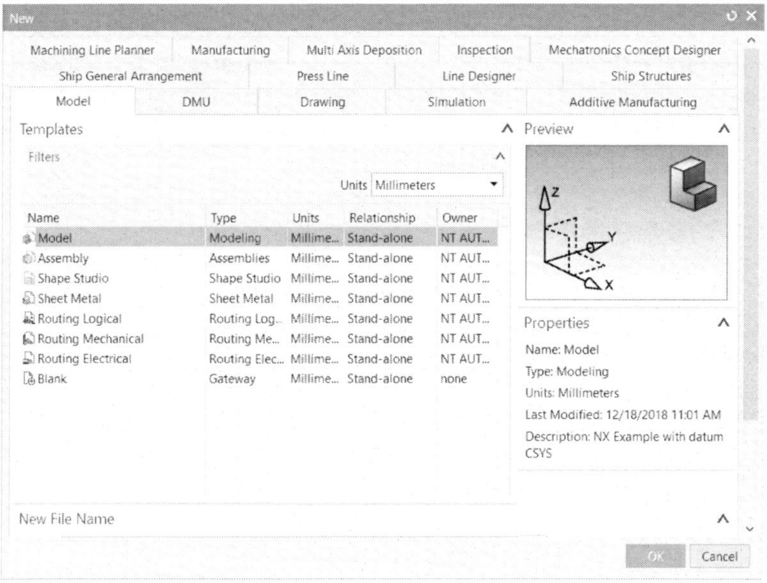

*Figure 1-2 The **New** dialog box*

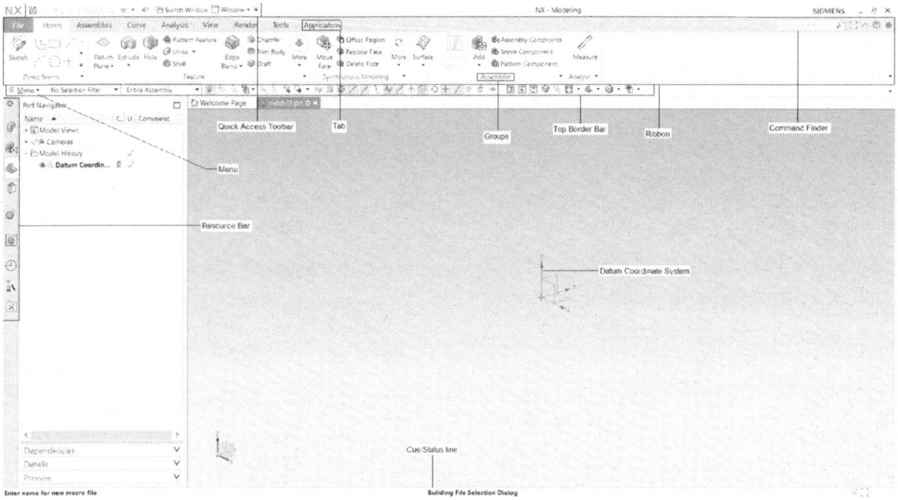

Figure 1-3 The Modeling environment

IMPORTANT TERMS AND DEFINITIONS

Some important terms and definitions of NX are discussed next.

Feature-based Modeling

A feature is defined as the smallest building block that can be modified individually. A model created in NX is a combination of a number of individual features and each feature is related to the other directly or indirectly. If a proper design intent is maintained while creating the model, then these features automatically adjust their values to any change in their surroundings. This provides a great flexibility to the design.

Parametric Modeling

The parametric nature of a software package is defined as its ability to use the standard properties or parameters in defining the shape and size of a geometry. The main function of this property is to derive the selected geometry to a new size or shape without considering its original dimensions. You can change or modify the shape and size of any feature at any stage of the designing process. This property makes the designing process an easy task. For example, consider the design of the body of a pipe housing, as shown in Figure 1-4.

To change the design by modifying the diameter of the holes and their number on the front, top, and bottom face, you need to select the feature and change the diameter and the number of instances in the pattern. The modified design is shown in Figure 1-5.

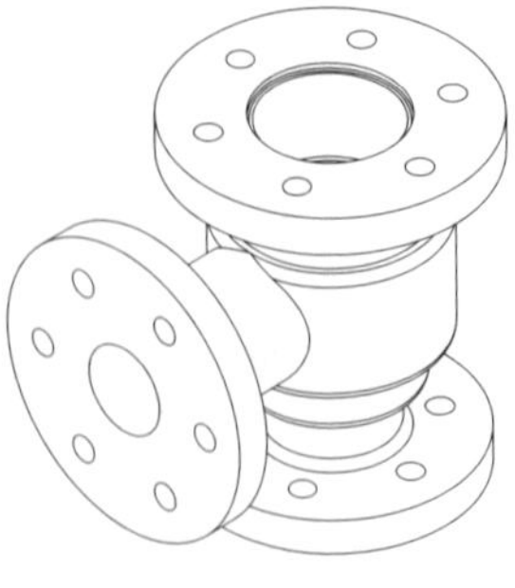

Figure 1-4 *Body of a pipe housing*

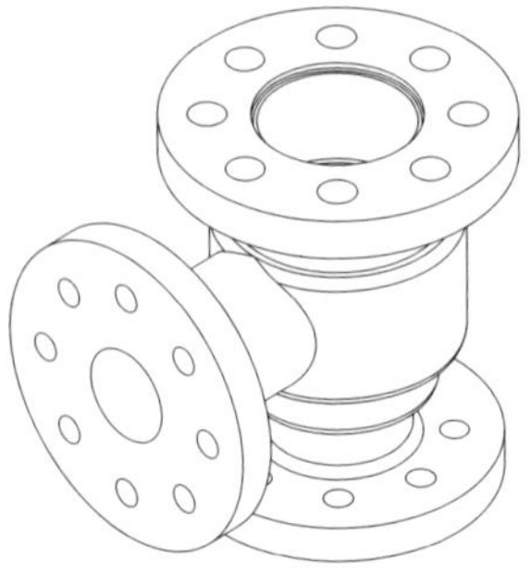

Figure 1-5 *Modified body of the pipe housing*

Chapter 2

Drawing Sketches for Solid Models

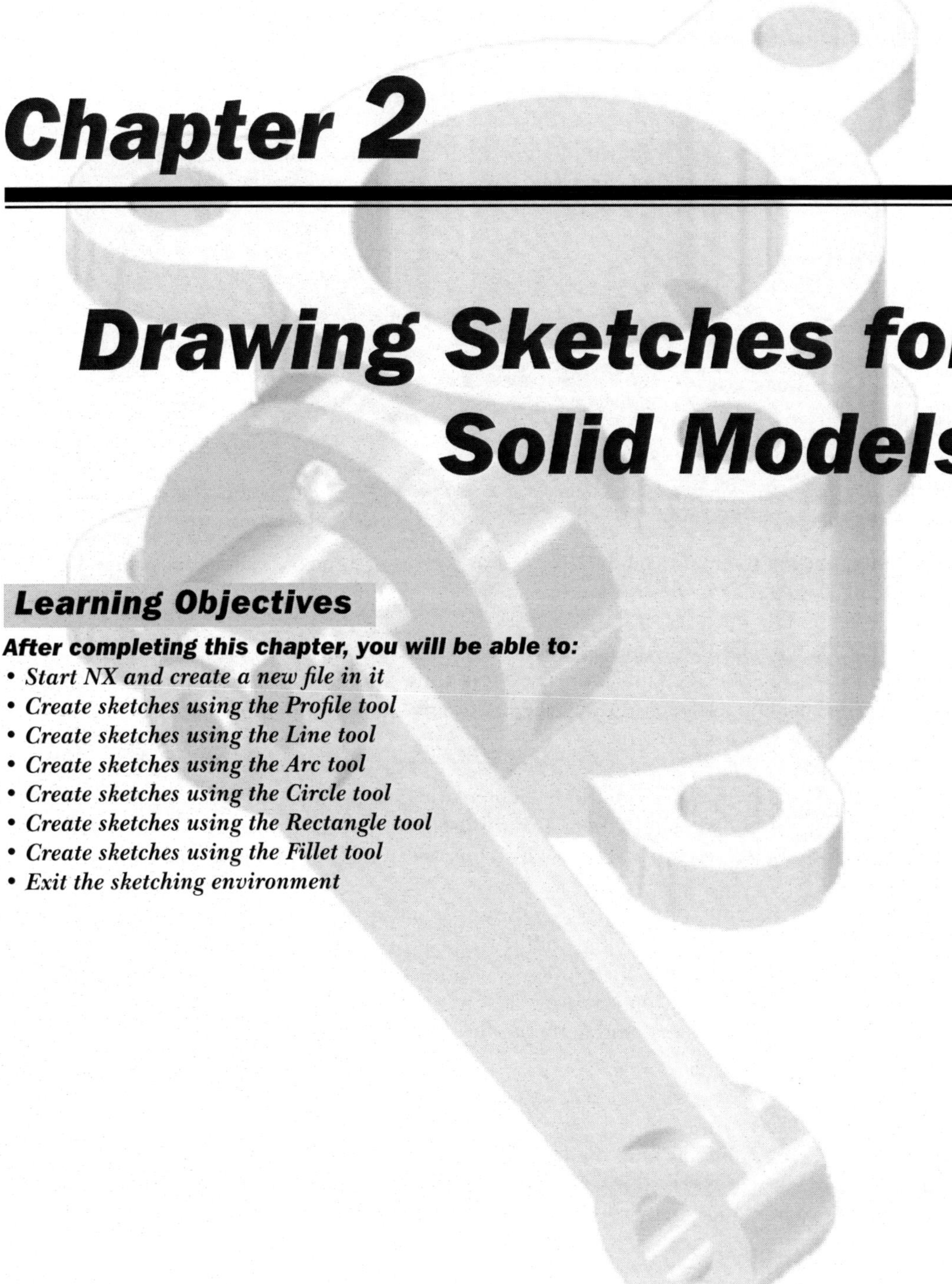

Learning Objectives

After completing this chapter, you will be able to:

- *Start NX and create a new file in it*
- *Create sketches using the Profile tool*
- *Create sketches using the Line tool*
- *Create sketches using the Arc tool*
- *Create sketches using the Circle tool*
- *Create sketches using the Rectangle tool*
- *Create sketches using the Fillet tool*
- *Exit the sketching environment*

INTRODUCTION

Most designs created in NX consist of sketch-based and placed features. A sketch is a combination of two-dimensional (2D) entities such as lines, arcs, circles, and so on. The features such as extrude, revolve, and sweep created by using 2D sketches are known as sketch-based features. The features such as fillet, chamfer, thread, and shell created without using a sketch are known as placed features. In a design, the base feature or the first feature is always a sketch-based feature. For example, the sketch shown in Figure 2-1 is used to create the solid model shown in Figure 2-2. In this model, the fillets and the chamfers are the placed features.

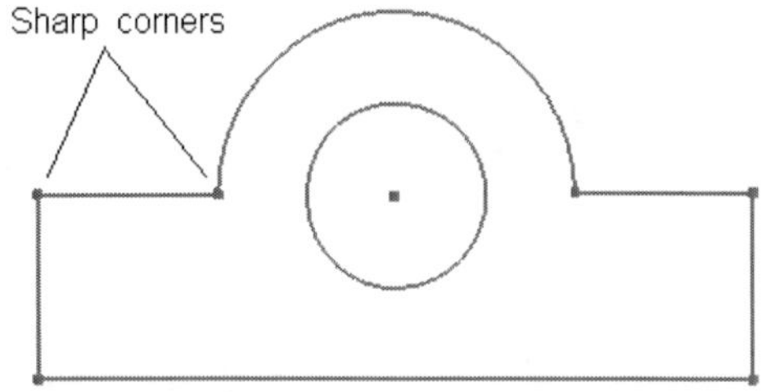

Figure 2-1 *Profile for the sketch-based feature of the solid model shown in Figure 2-2*

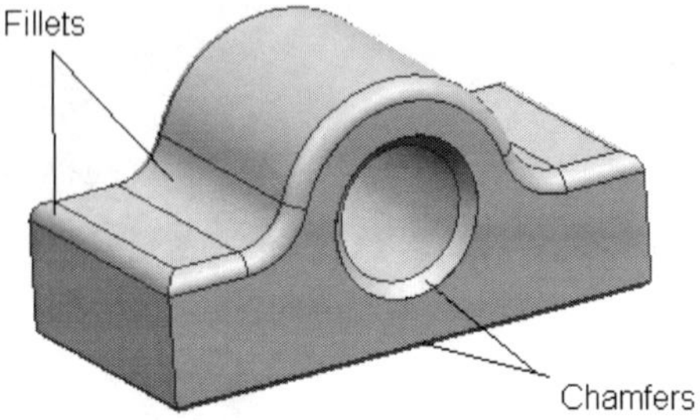

Figure 2-2 *Solid model created using the sketch-based and placed features*

As mentioned earlier, to create sketch-based features, you first need to create its sketch. In NX, you can create a sketch by using two methods: Direct sketch and Sketch in Task Environment.

In the Direct sketch method, you can create sketch, as required, directly in the Modeling environment by using the sketching tools such as **Line** and **Rectangle** of the **Direct Sketch** group. Once the sketch has been drawn, you can directly use the solid modeling tools to convert the sketch into a sketch-based feature.

In the Sketch in Task Environment method, to create the sketch, you need to invoke the sketching environment by using the **Sketch in Task Environment** tool from the **Curve** tab of the **Ribbon**. You will learn more about creating sketches by using these methods later in this chapter.

Unlike other solid modeling software packages where you need separate files for starting different environments, NX uses only a single type of file to start different environments. In NX, files are saved in the *.prt* format and all the environments required to complete a design can be invoked in the same *.prt* file. For example, you can draw sketches and convert them into features, assemble other parts with the current part, and generate drawing views in a single *.prt* file.

STARTING NX

You can start NX by double-clicking on its shortcut icon on the desktop of your computer. The default initial interface of NX is shown in Figure 2-3 and it displays basic information about NX. You can view more information by clicking on the topics displayed on the left of the NX screen.

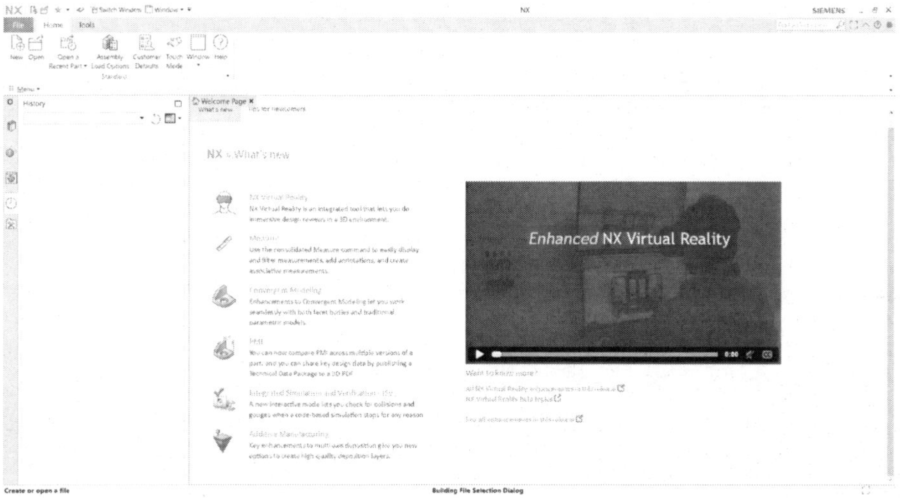

***Figure 2-3** The default initial interface of NX*

SKETCHING TOOLS

As discussed earlier, the tools required to draw a sketch are available in the **Sketch Curve** gallery of the **Direct Sketch** group in the **Home** tab of the Modeling environment. Also, these tools are available in the **Home** tab of the sketching environment. The sketching tools in context of the **Sketch Curve** gallery in the **Direct Sketch** group are discussed next.

Drawing Sketches Using the Profile Tool

Ribbon:	Home > Direct Sketch > Sketch Curve Gallery > Profile
Menu:	Insert > Sketch Curve > Profile

The **Profile** tool is the most commonly used tool to draw sketches in NX. This tool allows you to draw continuous lines and tangent/normal arcs. To draw continuous lines and tangent/normal arcs using this tool, choose the **Profile** tool from the **Direct Sketch** group of the **Home** tab; the **Profile** dialog box will be displayed, as shown in Figure 2-4.

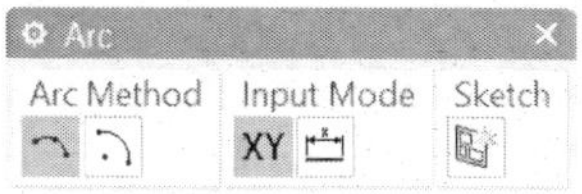

***Figure 2-4** The **Profile** dialog box*

Drawing Individual Lines

Ribbon:	Home > Direct Sketch > Sketch Curve Gallery > Line
Menu:	Insert > Sketch Curve > Line

NX also allows you to draw individual lines. This can be done by using the **Line** tool. The working of this tool is similar to the working of the line mode of the **Profile** tool. The only difference is that this tool allows you to draw only one line. As a result, after you specify the endpoint of the line, no rubber-band line is displayed. Instead, you will be prompted to specify the first point of the line. You can specify the first point and the second point of the lines by picking points on the screen or by entering values in the dynamic input boxes. You can use this tool to draw as many individual lines as required.

Drawing Arcs

Ribbon: Home > Direct Sketch > Sketch Curve Gallery > Arc
Menu: Insert > Sketch Curve > Arc

NX allows you to draw arcs using two methods. You can select a method by choosing its respective button from the **Arc** dialog box that will be displayed when you invoke the **Arc** tool.

Drawing Circles

Ribbon: Home > Direct Sketch > Sketch Curve Gallery > Circle
Menu: Insert > Sketch Curve > Circle

In NX, you can draw circles using two methods. These methods can be activated by choosing their respective buttons from the **Circle** dialog box that are displayed when you invoke the **Circle** tool.

Drawing Rectangles

Ribbon: Home > Direct Sketch > Sketch Curve Gallery > Rectangle
Menu: Insert > Sketch Curve > Rectangle

In NX, you can draw rectangles by using three methods. These methods can be used by choosing their respective buttons from the **Rectangle** dialog box. To invoke this dialog box, choose the **Rectangle** tool from the **Direct Sketch** group.

Filleting Sketched Entities

Ribbon: Home > Direct Sketch > Sketch Curve Gallery > Fillet
Menu: Insert > Sketch Curve > Fillet

Filleting is defined as the process of rounding the sharp corners of a profile to reduce the stress concentration. Fillets are created by removing the sharp corners and replacing them with round corners. In NX, you can create a fillet between any two sketched entities. This tool is available only in sketching environment.

EXITING THE SKETCH ENVIRONMENT

Ribbon: Home > Direct Sketch > Finish Sketch

Finish Sketch

After drawing the sketch, you need to exit the sketching environment to convert the sketch into a feature. To exit the sketching environment, choose the **Finish Sketch** tool from the **Direct Sketch** group of the **Home** tab in the **Ribbon**. Alternatively, right-click in the drawing area and choose the **Finish Sketch** option from the shortcut menu. When you exit the sketching environment, you can convert the sketch into a solid model by using the solid modeling tools. Note that you can also invoke solid modeling tools without exiting from the sketching environment.

TUTORIALS

As mentioned in the introduction, NX is parametric in nature. Therefore, you can draw a sketch of any dimensions and then modify its size by changing the values of dimensions. However, in this chapter, you will use the dynamic input boxes to draw the sketch of exact dimensions. This will help you improve your sketching skills.

Tutorial 1

In this tutorial, you will draw a profile for the base feature of the model shown in Figure 2-5. The profile to be drawn is shown in Figure 2-6. **(Expected time: 30 min)**

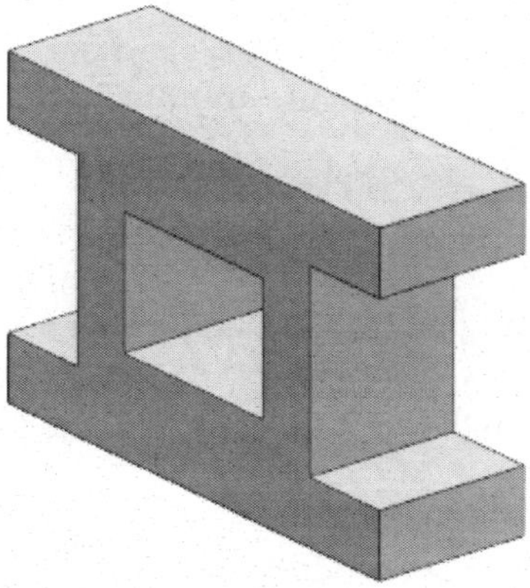

Figure 2-5 *Model for Tutorial 1*

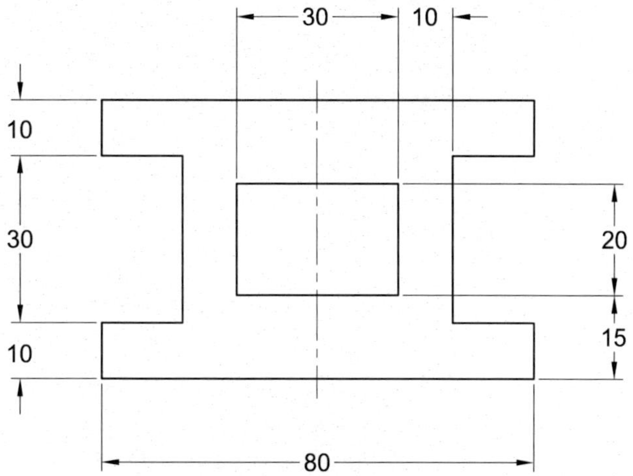

Figure 2-6 *Sketch for Tutorial 1*

The following steps are required to complete this tutorial:

a. Start a new file.
b. Select the XC-ZC plane as the sketching plane.
c. Draw the sketch of the model by using the **Profile** and **Rectangle** tools.
d. Finish the sketch and save the file.

Starting NX and Opening a New File

First, you need to start NX and then open a new file.

1. Double-click on NX shortcut icon on the desktop of your computer to start NX.

2. To start a new file, choose the **New** button from the **Standard** group of the **Home** tab or choose **Menu > File > New** from the **Top Border Bar**; the **New** dialog box is displayed.

3. Select the **Model** template from the **Templates** rollout.

4. Enter **c02tut1** as the name of the document in the **Name** text box of the dialog box.

5. Choose the button on the right of the **Folder** text box; the **Choose Directory** dialog box is displayed.

   It is recommended that you create a folder with the name NX in the hard drive of your computer and then create separate folders for each chapter inside it for saving the tutorial files of this textbook.

6. In this dialog box, browse to *NX /c02* folder and then choose the **OK** button twice; a new file is started in the Modeling environment.

Drawing the Sketch in the Modeling Environment

The base sketch of this model will be created on the XC-YC plane.

1. Choose the **Sketch** tool from the **Direct Sketch** group of the **Home** tab; the **Create Sketch** dialog box is displayed.

2. Select the XC-ZC plane from the Datum Coordinate system in the drawing window, if it is not selected by default.

3. Choose the **OK** button from the **Create Sketch** dialog box; the additional sketching tools become available and the sketching plane is oriented parallel to the screen.

Drawing the Outer Profile of the Sketch

The outer profile of the sketch consists of lines and it can be drawn by using the **Profile** tool.

1. Choose the **Profile** tool from the **Sketch Curve** gallery of the **Direct Sketch** group of the **Home** tab in the **Ribbon**; the **Profile** dialog box is displayed. By default, the **Line** button is active in this dialog box and the dynamic input boxes are displayed below the line cursor.

2. Move the cursor close to the origin; the coordinates of the point are displayed as **0,0** in the dynamic input boxes. Click to specify the start point of the line at this point.

   As you move the cursor on the screen, the line stretches and its length and angle values are modified dynamically in the dynamic input boxes.

3. Enter **80** in the **Length** dynamic input box and press the TAB key. Next, enter **0** in the **Angle** dynamic input box and press the ENTER key.

4. Choose the **Fit** tool from the **Operation** group of the **View** tab to fit the sketch into the drawing window.

5. Move the cursor away from the end point of the last line and then enter **10** as the length and **90** as the angle in the **Length** and **Angle** dynamic input boxes, respectively. Next, press the ENTER key.

6. Enter **15** as the length and **180** as the angle in the **Length** and **Angle** dynamic input boxes, respectively. Next, press the ENTER key.

7. Enter **30** as the length and **90** as the angle in the **Length** and **Angle** dynamic input boxes, respectively. Next, press the ENTER key.

8. Enter **15** as the length and **0** as the angle in the **Length** and **Angle** dynamic input boxes, respectively. Next, press the ENTER key.

9. Enter **10** as the length and **90** as the angle in the **Length** and **Angle** dynamic input boxes, respectively. Next, press the ENTER key.

10. Enter **80** as the length and **180** as the angle in the **Length** and **Angle** dynamic input boxes, respectively. Next, press the ENTER key.

11. Enter **10** as the length and **-90** or **270** as the angle in the **Length** and **Angle** dynamic input boxes, respectively. Next, press the ENTER key.

12. Enter **15** as the length and **0** as the angle in the **Length** and **Angle** dynamic input boxes, respectively. Next, press the ENTER key.

13. Enter **30** as the length and **-90** or **270** as the angle in the **Length** and **Angle** dynamic input boxes, respectively. Next, press the ENTER key.

14. Enter **15** as the length and **180** as the angle in the **Length** and **Angle** dynamic input boxes, respectively. Next, press the ENTER key.

15. Enter **10** as the length and **-90** or **270** as the angle in the **Length** and **Angle** dynamic input boxes, respectively. Next, press the ENTER key.

16. Press the ESC key twice to exit the **Profile** tool. The outer profile of the sketch is shown in Figure 2-7.

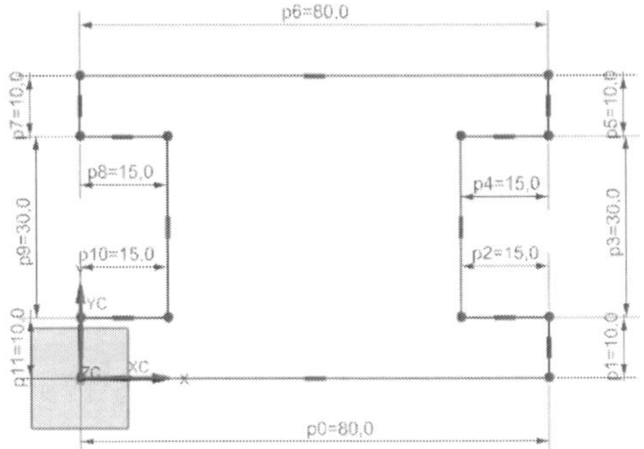

Figure 2-7 *Outer profile of the sketch*

Drawing the Rectangle

Next, you need to draw the inner profile, which is a rectangle. You can use the **By 2 Points** option of the **Rectangle** tool to draw the rectangle.

1. Choose the **Rectangle** tool from the **Sketch Curve** gallery of the **Direct Sketch** group; the **Rectangle** dialog box is displayed and the **By 2 Points** button is chosen by default in this dialog box.

2. Enter **25** and **15** as the coordinates of the first point of the rectangle in the **XC** and **YC** dynamic input boxes, respectively. Next, press the ENTER key.

3. Enter **30** and **20** as the width and height of the rectangle in the **Width** and **Height** dynamic input boxes, respectively. Next, press the ENTER key. A preview of the rectangle is displayed, but it is not actually drawn yet. As you move the cursor in the drawing window, the rectangle also moves.

4. Move the cursor close to the top right corner of the drawing window and then click to draw the rectangle.

5. Press the ESC key to exit the tool. The final sketch for Tutorial 1 is shown in Figure 2-8.

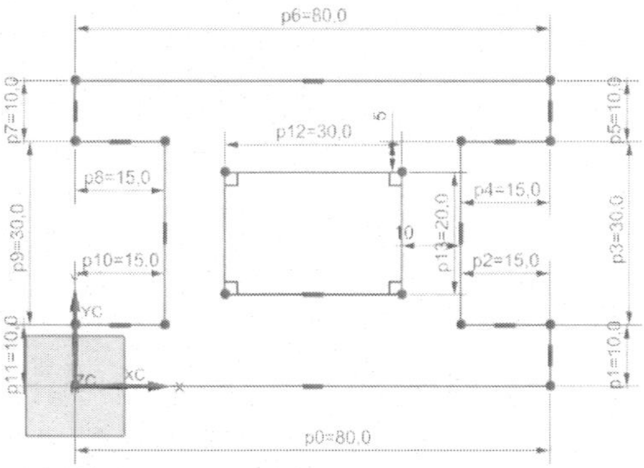

Figure 2-8 *Final sketch for Tutorial 1*

Finishing the Sketch and Saving the File

NX allows you to save the sketch file in the Sketch environment.

1. Choose the **Save** button from the **Quick Access** toolbar.

2. Next, choose the **Finish Sketch** tool from the **Direct Sketch** group; the Modeling environment is invoked.

3. Choose **Menu > File > Close > Selected Parts** from the **Top Border Bar**; the **Close Part** dialog box is displayed.

4. Select the name of the current file from the list area in the **Part** rollout and then choose the **OK** button to close the current file.

Tutorial 2

In this tutorial, you will draw the profile of the model shown in Figure 2-9. The profile to be drawn is shown in Figure 2-10. **(Expected time: 30 min)**

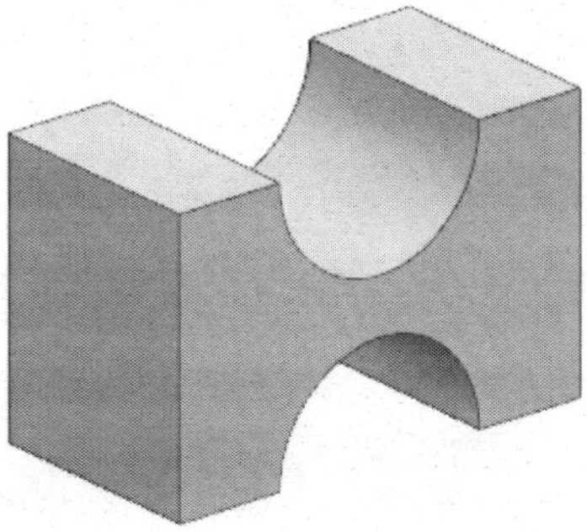

Figure 2-9 *Model for Tutorial 3*

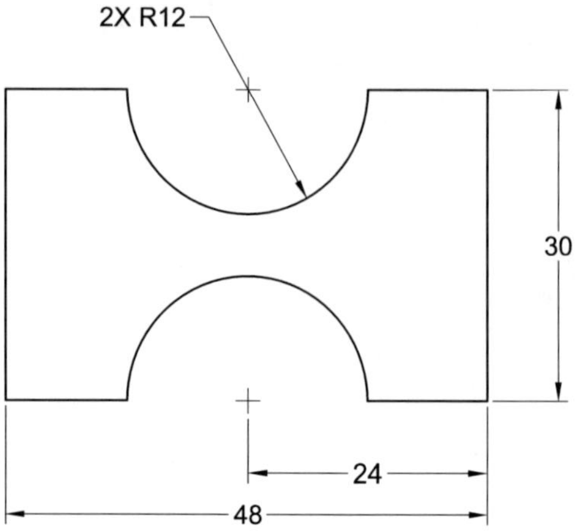

Figure 2-10 *Sketch for Tutorial 3*

The following steps are required to complete this tutorial:

a. Start a new file.
b. Select the YC-ZC plane as the sketching plane.
c. Draw the sketch of the model by using the **Profile** tool.
d. Finish the sketch and save the file.

Starting a New File

If you continue working after completing Tutorial 1, you do not need to open a new session of NX. You can start a new part file by selecting the **Model** template from the **New** dialog box.

1. To start a new file, choose the **New** button from the **Standard** group of the **Home** tab or choose **Menu > File > New** from the **Top Border Bar**; the **New** dialog box is displayed.

2. Select the **Model** template from the **Templates** rollout.

3. Choose the button on the right of the **Name** text box; the **Choose New File Name** dialog box is displayed.

4. In this dialog box, browse to *NX/c02* and then enter **c02tut2** in the **File name** edit box. Next, choose the **OK** button twice; the new file is started in the Modeling environment.

Drawing the Sketch in the Modeling Environment

The base sketch of this model will be created on the YC-ZC plane. Therefore, you need to draw the sketch using this plane.

1. Choose the **Sketch** tool from the **Direct Sketch** group; the **Create Sketch** dialog box is displayed.

2. Select the YC-ZC plane from the drawing window. Note that the Z-axis direction of the sketching plane points toward the front side of the sketching plane and the direction of Y-axis is upward.

3. Choose the **OK** button from the **Create Sketch** dialog box to start the sketch.

Drawing the Sketch

The sketch that you need to draw consists of multiple lines and two arcs. All these entities can be drawn by using the **Line** and **Arc** options of the **Profile** tool.

1. Choose the **Profile** tool from the **Sketch Curve** gallery of the **Direct Sketch** group; the **Profile** dialog box is displayed. In this dialog box, the **Line** button is activated by default. Also, the dynamic input boxes are displayed below the line cursor.

2. Move the cursor close to the origin; the coordinates of the point are displayed as 0,0 in the dynamic input boxes. Click to specify the start point of the line at this point.

   The point you specified is selected as the start point of the line and the endpoint is attached to the cursor. As you move the cursor on the screen, the line stretches and its length and angle values are modified dynamically in the dynamic input boxes.

   Next, you need to specify the endpoint of this line. Also, you need to specify the points to define the remaining lines of the sketch. This can be done by using the **Length** and **Angle** dynamic input boxes.

3. Enter **12** in the **Length** dynamic input box and press the TAB key. Next, enter **0** in the **Angle** dynamic input box and press the ENTER key. The first line is drawn and a rubber-band line is displayed with the start point at the endpoint of the previous line and the endpoint attached to the cursor.

   Now, you need to invoke the arc mode because the next entity to be drawn is an arc.

4. Choose the **Arc** button from the **Object Type** area of the **Profile** dialog box to invoke the arc mode.

   A rubber-band arc is displayed with the start point fixed at the endpoint of the last line and the endpoint attached to the cursor. Also, the quadrant symbol is displayed at the start point of the arc.

5. Move the cursor to the start point of the arc and then move it vertically upward through a small distance. Next, move the cursor toward the right; you will notice that a normal arc starts from the endpoint of the last line.

6. Enter **12** in the **Radius** dynamic input box and press the TAB key. Next, enter **180** in the **Sweep Angle** dynamic input box and press the ENTER key.

   A preview of the resulting arc is displayed, but the arc is still not drawn. To draw the arc, you need to specify a point on the screen with the values mentioned in the dynamic input boxes.

7. Move the cursor horizontally toward the right and click when the preview of the required arc is displayed. The arc is drawn and the line mode is invoked again.

8. Enter **12** as the length and **0** as the angle in the **Length** and **Angle** dynamic input boxes, respectively, and then press the ENTER key. Choose the **Fit** tool from the **Operation** group of the **View** tab to fit the sketch into the drawing window.

9. Enter **30** as the length and **90** as the angle in the **Length** and **Angle** dynamic input boxes, respectively, and then press the ENTER key.

10. Move the cursor horizontally toward the left. Make sure that the horizontal constraint symbol is displayed. Click to specify the endpoint of the line when the vertical help line is displayed from the endpoint of the arc.

    Next, you need to draw the arc by invoking the arc mode.

11. Choose the **Arc** button from the **Profile** dialog box to invoke the arc mode; a rubber-band arc is displayed with its start point fixed at the endpoint of the last line.

12. Move the cursor to the start point of the arc and then move it vertically downward through a small distance. When the normal arc appears, move the cursor toward the left.

13. Move the cursor over the lower arc once and then move it toward the left, refer to Figure 2-11.

    A horizontal help line is displayed originating from the center of the arc being drawn. At the point where the cursor is vertically in line with the start point of the lower arc, a vertical help line appears from the start point of the lower arc, refer to Figure 2-11.

14. Click to define the endpoint of the arc when the horizontal and vertical help lines are displayed. The arc is drawn and the line mode is invoked again.

15. Enter **12** as the radius and **180** as the angle in the **Length** and **Angle** dynamic input boxes, respectively, and then press the ENTER key.

16. Move the cursor to the first line and then move it to the start point of this line; the cursor snaps to the start point of the line.

17. Click to define the endpoint of this line when the cursor snaps to the start point of the first line.

18. Press the ESC key twice to exit the **Profile** tool. The final sketch of the model is shown in Figure 2-12.

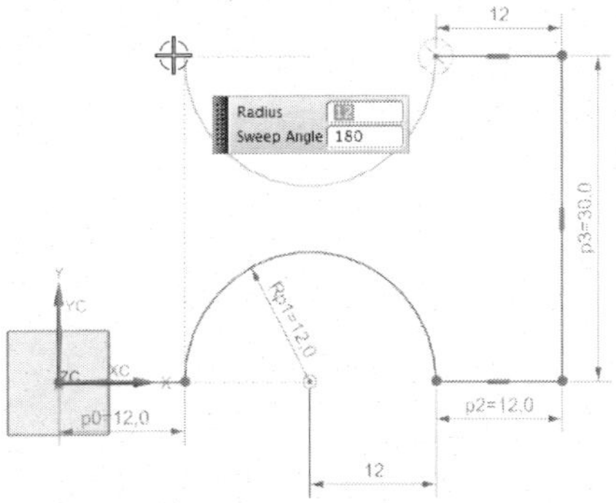

Figure 2-11 Horizontal and vertical help lines displayed to define the endpoint of the arc

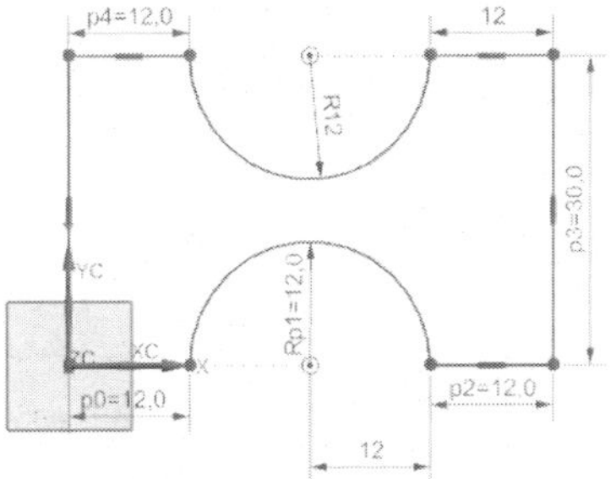

Figure 2-12 Final sketch for Tutorial 3

Finishing the Sketch and Saving the File

NX allows you to save the sketch file in the Sketch environment.

1. Choose the **Save** button from the **Quick access** toolbar or from the **File** tab to save the sketch.

2. Next, choose the **Finish Sketch** tool from the **Direct Sketch** group; the Modeling environment is invoked.

3. Choose **Menu > File > Close > Selected Parts** from the **Top Border Bar**; the **Close Part** dialog box is displayed.

4. Select the name of the current file from the list area in the **Part** rollout and then choose the **OK** button to close the current file.

EXERCISES

Exercise 1

Draw a sketch for the base feature of the model shown in Figure 2-13. The sketch to be drawn is shown in Figure 2-14. **(Expected time: 30 min)**

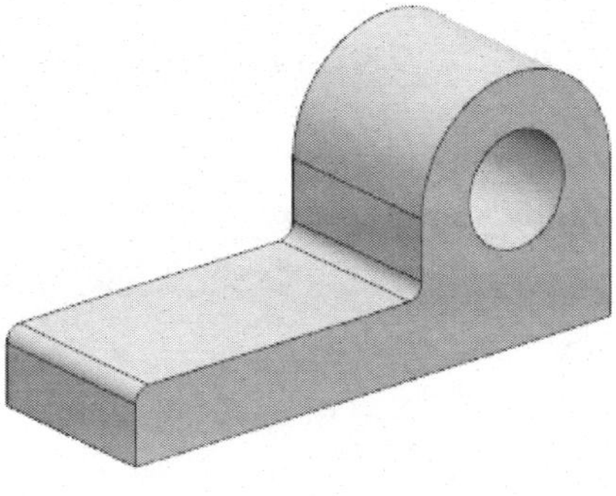

Figure 2-13 Model for Exercise 1

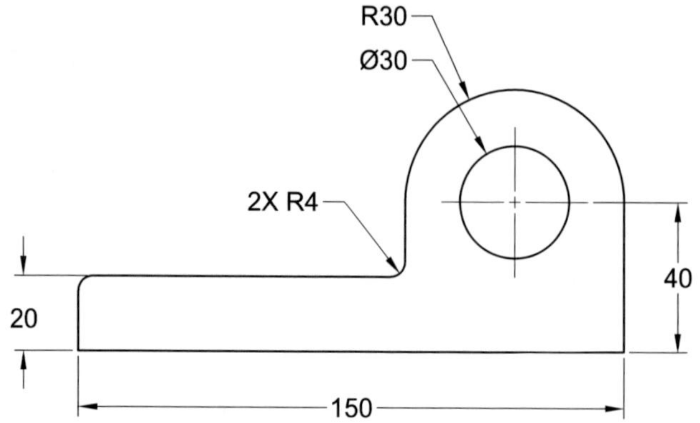

Figure 2-14 Sketch for Exercise 1

Exercise 2

Draw a sketch for the base feature of the model shown in Figure 2-15. The sketch to be drawn is shown in Figure 2-16. **(Expected time: 30 min)**

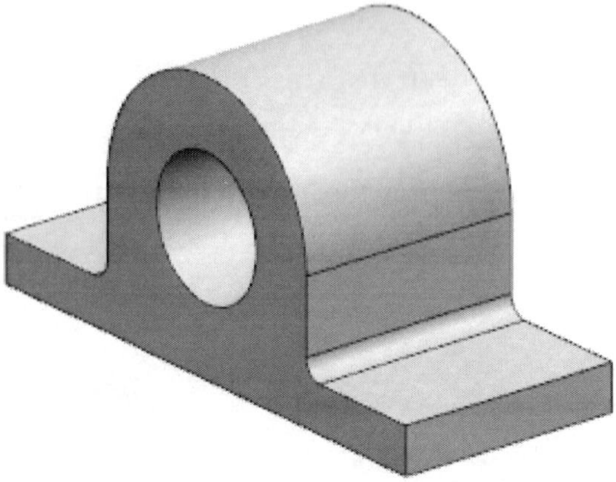

Figure 2-15 Model for Exercise 2

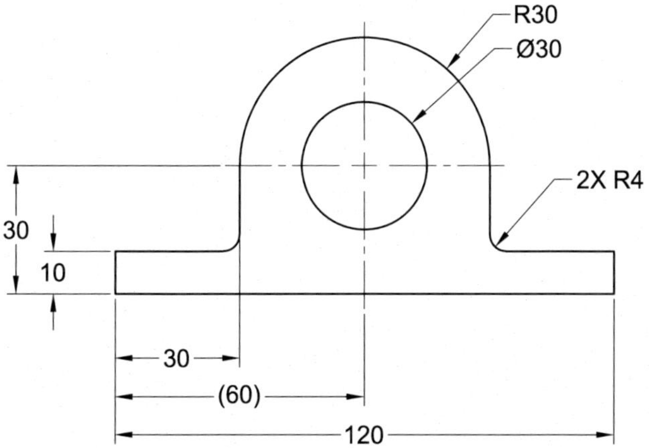

Figure 2-16 Sketch for Exercise 2

This page is intentionally left blank

Chapter 3

Adding Geometric and Dimensional Constraints to Sketches

Learning Objectives

After completing this chapter, you will be able to:

- *Understand the concept of under-constrained, fully-constrained, and over-constrained sketches*
- *Understand different types of dimensions*
- *Understand different types of geometric constraints*
- *Understand to show sketch constraints*

CONSTRAINING SKETCHES

In the previous chapter, you learned to draw sketches. You must have noticed that the dimensions are applied automatically to the sketches when they are drawn but the dimensions are updated when you drag the sketched entities. Therefore, these dimensions do not constrain the sketch. In this chapter, you will learn to completely constrain the sketches to restrict their degrees of freedom and make them stable. The stability ensures that the size, shape, and location of the sketches do not change unexpectedly with respect to the surrounding. Therefore, it is always recommended to constrain the sketches. The geometrical constraints are applied first, some of which get automatically applied while drawing. After applying the remaining geometrical constraints, you need to add dimensional constraints using dimension tools. You will learn more about dimension tools later in this chapter.

CONCEPT OF CONSTRAINED SKETCHES

After drawing and applying the constraints, the sketch can attain any one of the following three stages:

1. Under-Constrain
2. Fully-Constrain
3. Over-Constrain

These stages are described next.

Under-Constrain

An under-constrained sketch is the one in which all the degrees of freedom of each entity are not completely defined using the geometric and dimensional constraints. The elements of the sketch that are displayed in light purple color are under-constrained. You need to apply additional constraints to them in order to constrain their degree of freedom. The under-constrained sketches tend to change their position, size, or shape unexpectedly. Therefore, it is necessary to fully define the sketched elements. Figure 3-1 shows an under-constrained sketch.

Fully-Constrain

The fully-constrained sketch is the one in which all degrees of freedom of each element are defined using the geometric and dimensional constraints. As a result, the sketch cannot change its position, shape, or size unexpectedly. These dimensions can change only if they are modified deliberately by the user. The elements of a fully-constrained sketch are displayed in dark green color. Figure 3-2 shows a fully-constrained sketch.

Over-Constrain

An over-constrained sketch is the one in which some additional constraints are applied. The over-constrained dimensions and constraints are displayed in red color. It is always recommended to delete additional constraints and make the sketch fully-constrained before exiting the sketching environment. Figure 3-3 shows an over-constrained sketch.

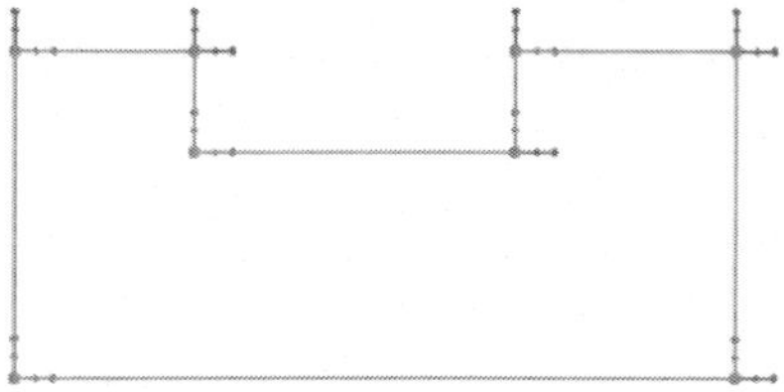

Figure 3-1 *An under-constrained sketch*

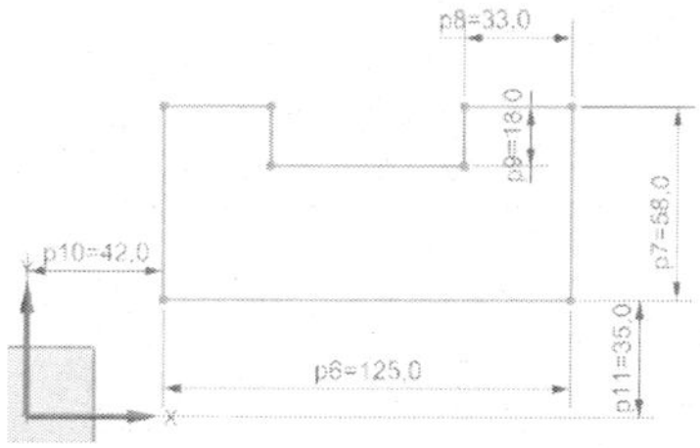

Figure 3-2 *A fully-constrained sketch*

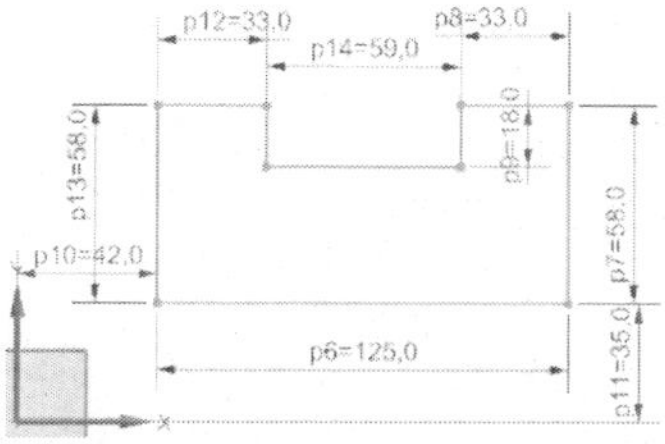

Figure 3-3 *An over-constrained sketch*

While applying the geometric and dimensional constraints, the status area in the Status Bar displays the number of constraints needed to fully constrain the sketch. After fully constraining the sketch, a message will be displayed in the status area of the **Status Bar** that the sketch is fully constrained.

Also, if the sketch is over-constrained, you will be informed that the sketch contains over constrained geometry. In such a case, you need to remove the unwanted constraints.

DIMENSIONING SKETCHES

After creating a sketch, you need to apply different types of dimensions (dimensional constraints) to it. The purpose of dimensioning is to control the size of the sketch and to place it with reference to some other entity. As discussed earlier, by default, sketched entities get dimensioned automatically while drawing a sketch. However, the automatically applied dimensions do not constrain the sketch. Therefore, you need to lock them to fully constrain the sketch. Also, sometimes you may need to add additional dimensions to the sketch to fully constrain it. In NX, you can apply the dimensions by using the tools grouped together in the **Dimensions** Drop-down of the **Direct Sketch** group in the **Home** tab, refer to Figure 3-4. You can also access these tools from the **Dimensions** Drop-down of the **Constraints** group in the **Home** tab of the Sketch in Task Environment. The tools available in this drop-down are listed below:

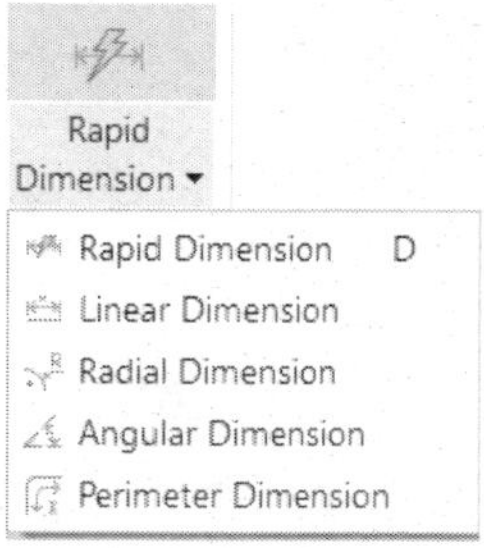

Figure 3-4 *The* ***Dimensions Drop-down***

1. Rapid Dimension
2. Linear Dimension
3. Radial Dimension

4. Angular Dimension
5. Perimeter Dimension

GEOMETRIC CONSTRAINTS

Geometric constraints are the logical operations that are performed on the sketched entities to relate them to the other sketched entities using the standard properties such as collinearity, concentricity, tangency, and so on. These constraints reduce the degrees of freedom of the sketched entities and make the sketch more stable so that it does not change its shape and location unexpectedly at any stage of the design. All geometric constraints have separate symbols associated with them. These symbols can be seen on the sketched entities when the constraints are applied to them. In the sketching environment of NX, you can add various types of geometric constraints. Some of these constraints are added automatically while sketching. Additionally, you can add more constraints to the sketch manually. These are explained next.

Applying Additional Constraints Individually

Ribbon:	Home > Direct Sketch > More Gallery > Sketch Constraints > Geometric Constraints
Menu:	Insert > Sketch Constraint > Geometric Constraints

In NX, you can apply additional constraints manually by using the **Geometric Constraints** tool in the **Direct Sketch** group of the **Home** tab. To apply constraints, invoke the **Geometric Constraints** tool; the **Geometric Constraints** message box will be displayed. Select the **Don't display this message again** check box in it and then choose the **OK** button; the **Geometric Constraints** dialog box will be displayed, as shown in Figure 3-5. Select the required constraint from the **Constraint** rollout of the dialog box and then select the entities from the drawing area to apply the selected constraint.

Note

*By default, only few constraints are available in the **Constraint** rollout of the **Geometric Constraints** dialog box. You can customize constraints to add or remove the constraints from the **Constraint** rollout. Select the check boxes next to the required constraints from the **Settings** rollout; the selected constraints will be displayed in the **Constraint** rollout.*

In NX, you can also apply constraints directly without invoking the **Geometric Constraints** dialog box. To do so, select the entities from the drawing area; a contextual toolbar will be displayed with all possible constraints that can be applied to the selected entities, as shown in Figure 3-6. Select the required constraint from the contextual toolbar; the selected constraint will be applied to the selected entities.

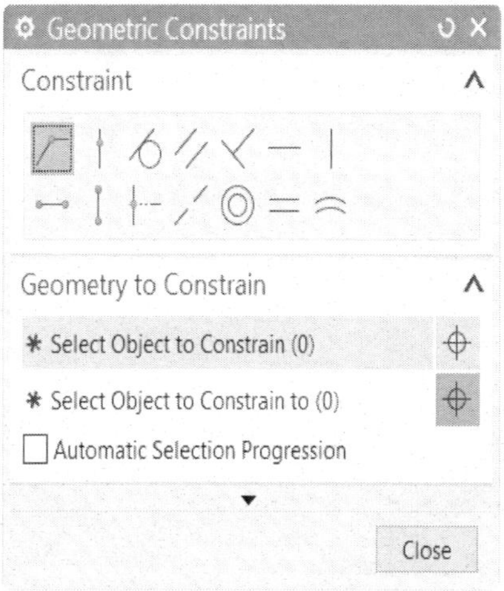

Figure 3-5 *The* ***Geometric Constraints*** *dialog box*

Figure 3-6 *The contextual toolbar*

Showing All Constraints in a Sketch

Ribbon:	Home > Direct Sketch > More Gallery > Sketch Constraints > Display Sketch Constraints
Menu:	Tools > Sketch Constraints > Display Sketch Constraints

By default, all the constraints that are applied to a sketch are displayed in the sketch. You can turn on/off the display of these constraints by using the **Display Sketch Constraints** button. This is a toggle button and when you turn it off, it remains off until you turn it on. With this button turned on, you can continue working with the other sketching tools.

TUTORIALS

From this chapter onward, you will use Sketch constraints and parametric dimensions to complete the model.

Tutorial 1

In this tutorial, you will draw the profile of the model shown in Figure 3-7. The profile is shown in Figure 3-8. The profile should be symmetric about the origin. Also, you will use the parametric dimensions to complete the sketch. **(Expected time: 30 min)**

Figure 3-7 *Model for Tutorial 1*

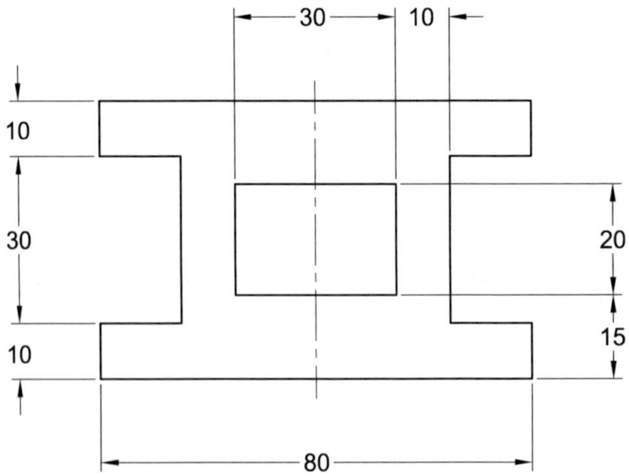

Figure 3-8 *Sketch for Tutorial 1*

The following steps are required to complete this tutorial:

a. Start a new file.
b. Draw the outer profile of the sketch using the **Profile** tool, refer to Figure 3-9.
c. Add the geometric constraints to the outer loop and modify its dimensional constraints, refer to Figure 3-10 and 3-11.
d. Draw a rectangle inside the outer loop using the **Rectangle** tool, refer to Figure 3-12.
e. Add dimensions to the rectangle to complete the sketch, refer to Figure 3-13.
f. Save the sketch and close the file.

Starting a New File and Invoking the Sketch Environment

1. Start a new file by using the **Model** template. To do so, choose the **New** button from the **File** tab of the **Ribbon**; the **New** dialog box is displayed. Next, select the **Model** template from the **Templates** rollout, and then enter **c03tut1** as the name of the document in the **Name** text box.

2. Choose the button on the right of the **Folder** text box; the **Choose Directory** dialog box is displayed. Next, browse to *C:\NX\c03* and then choose the **OK** button twice; the new file is started in the Modeling environment.

3. Choose the **Sketch** tool from the **Direct Sketch** group; the **Create Sketch** dialog box is displayed. Select the XC-ZC plane as the sketching plane and choose the **OK** button from the dialog box.

Note

*You can also invoke the Sketch environment by choosing the **Sketch in Task Environment** tool from the **Curve** tab of the **Ribbon** and create sketch.*

Drawing the Outer Loop and Adding Sketch Constraints

If the sketch consists of more than one closed loop, it is recommended that you draw the outer loop first and then add all the required Sketch constraints and dimensions to it. This makes it easier to draw and dimension the inner loops. Next, you need to draw the inner loop.

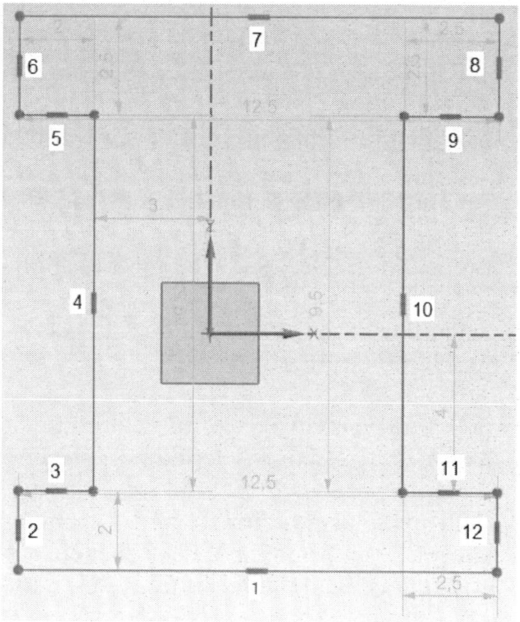

***Figure 3-9** The outer loop of the profile and the sequence in which lines have to be drawn*

1. Choose the **Profile** tool from the **Direct Sketch** group of the **Home** tab; you are prompted to select the first point of the line or press and drag the left mouse button to begin with the arc creation.

2. Draw the sketch around the origin following the sequence shown in Figure 3-9. The sketch is unsymmetrical at this stage. But after adding the Sketch constraints and modifying the dimensions, it will become symmetrical. You can use the help lines to draw the sketch. For your reference, the sequence in which the lines are needed to be drawn in the sketch are indicated by numbers.

Note

The dimensions shown in Figure 3-9 get automatically applied when sketch is drawn. However, those dimensions are not driving dimensions. They will change when you drag or move a sketched entity. In order to make them driving dimensions, you need to modify them. You will learn to modify these dimensions later in this tutorial.

Next, you need to apply the geometric constraints and modify dimensions. But before you do that, it is recommended that you turn on the display of constraints, if it is not already on.

3. Choose the **Display Sketch Constraints** tool from the **Sketch Constraints** rollout of the **More** gallery in the **Direct Sketch** group to view the constraints applied to the sketch, if it is not chosen by default.

4. Choose the **Geometric Constraints** tool from the **More** gallery of the **Direct Sketch** group of the **Home** tab; the **Geometric Constraints** dialog box is displayed and you are prompted to select curves to create constraints.

5. Choose the **Equal Length** button from the **Geometric Constraints** dialog box and then select lines 1 and 7. The symbol for the equal length constraint is displayed on both the sketch members indicating that this constraint is applied between the two selected entities. Note that **Automatic Selection Progression** check box should be selected in the **Geometry to Constrain** rollout to move to the succeeding points.

6. Similarly, apply the equal length constraint between lines 8 and 6, 6 and 2, 2 and 12, 12 and 8, 3 and 11, 9 and 5, and 10 and 4. Next, exit the dialog box.

   Now, you need to make this sketch symmetric about horizontal and vertical datum axes.

7. Choose the **Make Symmetric** tool from the **Sketch Constraints** rollout of the **More** gallery in the **Direct Sketch** group; the **Make Symmetric** dialog box is displayed and you are prompted to select the first object to apply symmetry.

8. Select line 2; you are prompted to select the second object. Select line 12; you are prompted to select the centerline.

9. Select the vertical axis as the centerline; line 2 and 12 are made symmetric about vertical axis. if vertical axis is selected as centerline, choose the **Reset** button from the dialog box

10. Similarly, make the lines 1 and 7 symmetric about the horizontal axis. Next, choose the **Close** button from the **Make Symmetric** dialog box to close it. The sketch after applying constraints to all these entities is shown in Figure 3-10.

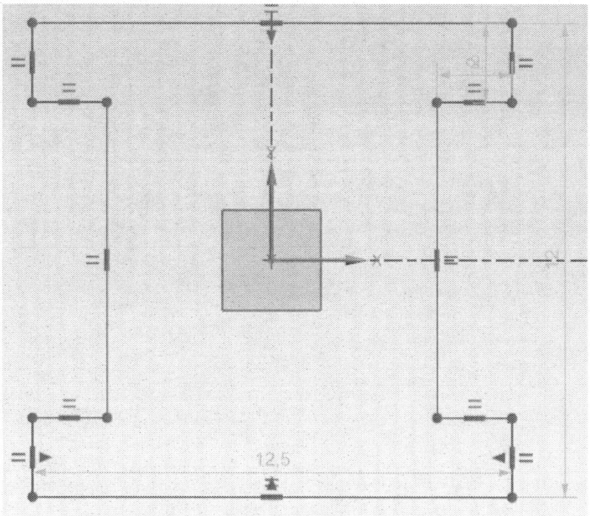

Figure 3-10 The outer profile after adding constraints

Adding Dimensions to Sketch Members

Next, you need to add dimensions to the sketch. As mentioned earlier, when you add dimensions to a sketch and modify their values, the entity is forced by the specified dimension value to maintain the modification. Before you start dimensioning the sketch, you need to modify some dimension display options.

1. Choose the **Rapid Dimension** tool from the **Direct Sketch** group of the **Home** tab; you are prompted to select an object to dimension or the dimension to edit.

2. Select line 1; the current dimension of line 1 isattached to the cursor. Now, you need to place the dimension at the required location. Click below the line 1 to place the dimension. As you place the dimension, an edit box is displayed. Enter **80** in the edit box and press ENTER, refer to Figure 3-11. Next, choose the **Fit** tool from the **Operation** group of the **View** tab.

3. Select line 2 and place the dimensions on the left of the sketch. Enter **10** in the edit box displayed and press ENTER.

4. Select line 4 and place the dimension on the left of the sketch. Next, modify the dimension value to **30** and press ENTER.

5. Select line 9 and place the dimension below the line. Next, modify the dimension value to **15** and press ENTER.

   When you place the dimensions, they generally scatter all around the sketch. It is a good practice to arrange them properly.

6. Exit the **Rapid Dimension** tool by pressing the ESC key and then drag the dimensions to place them properly around the sketch, refer to Figure 3-11.

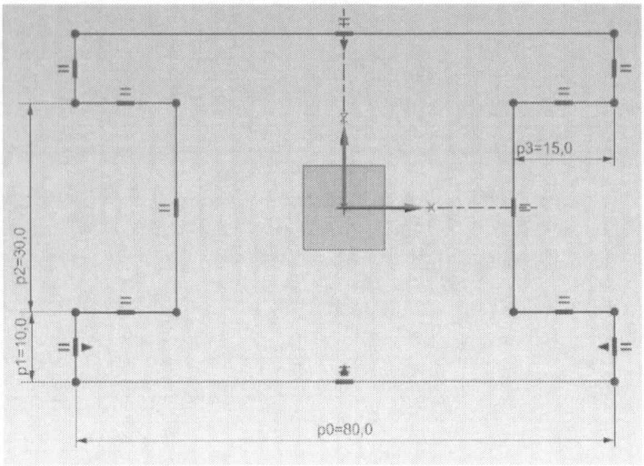

***Figure 3-11** The outer profile after adding the required dimensions*

> **Tip**
> *Instead of adding dimensions to a sketch, you can also modify the existing dimension. To do so, double-click on the dimension; an edit box will be displayed. Enter the required value in the edit box and press ENTER; the dimensions will be locked and displayed in blue color.*

Drawing the Inner Loop and Adding Dimensions

Next, you need to draw the rectangular profile inside the outer loop.

1. Choose the **Rectangle** tool from the **Direct Sketch** group of the **Home** tab; the **Rectangle** dialog box is displayed.

2. Choose the **From Center** button from the **Rectangle** dialog box; you are prompted to specify the center of the rectangle.

3. Select origin as the center of the rectangle; you are prompted to specify the second point of the rectangle.

4. Move the cursor horizontally toward right and click to specify the second point; you are prompted to select the third point to create the rectangle.

5. Move the cursor vertically upward and click to specify the third point; the rectangle is created, refer to Figure 3-12. Press ESC to exit the tool.

6. Double-click on the horizontal dimension of the rectangle; a dynamic edit box is displayed. Modify the value of the dimension to **30** and press ENTER. Next, drag and place the dimension above the sketch.

7. Similarly, double-click on the vertical dimension of the rectangle and modify the value of the dimension to **20**. Next, press ESC. Next, place the dimension on the left of the sketch, refer to Figure 3-13.

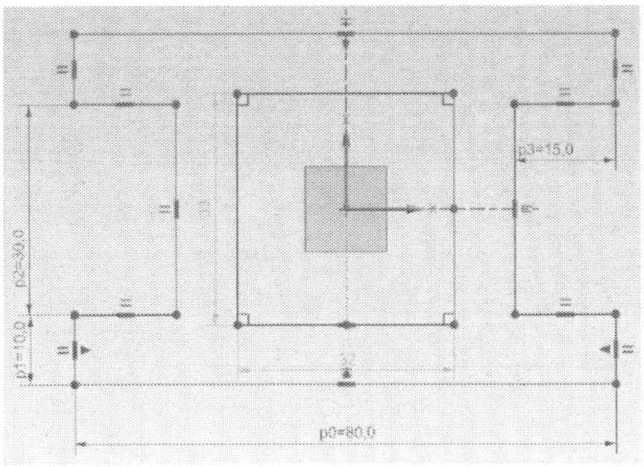

Figure 3-12 *The sketch after drawing the inner loop and turning on the display of constraints*

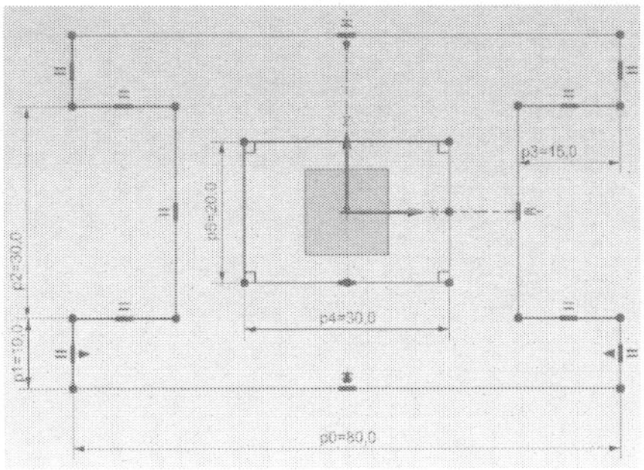

Figure 3-13 *The sketch after adding the required dimensions and constraints*

Saving and Closing the File

1. Exit the sketching environment by choosing the **Finish Sketch** tool from the **Direct Sketch** group of the **Home** tab. Next, choose the **Save** button from the **Quick Access** toolbar to save the sketch. Note that the name and location of the document had already been specified when you started new file.

2. Choose **Menu > File > Close > All Parts** from the **Top Border Bar** to close the file.

Tutorial 2

In this tutorial, you will create the profile of the revolved model shown in Figure 3-14. The profile is shown in Figure 3-15. You will use the geometric and dimensional constraints to complete this sketch. **(Expected time: 30 min)**

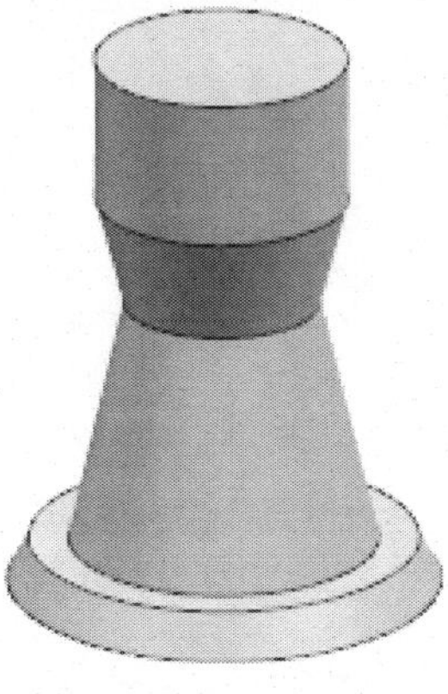

Figure 3-14 *Model for Tutorial 3*

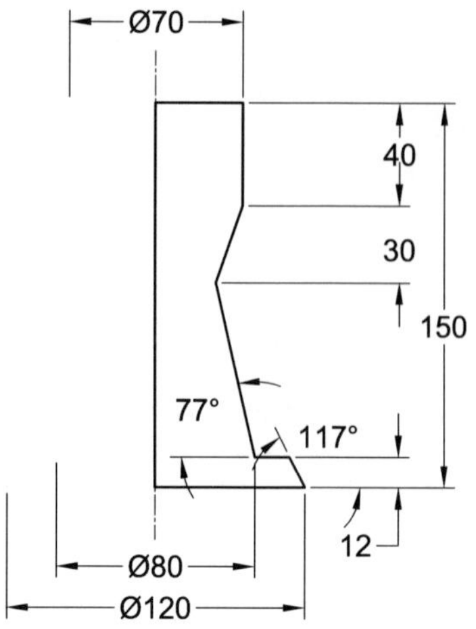

Figure 3-15 *Sketch for Tutorial 3*

The following steps are required to complete this tutorial:

a. Start a new file in NX and invoke the Sketch environment.
b. Draw the required profile of the sketch using the **Profile** tool, refer to Figure 3-16.
c. Add the geometric and dimensional constraints to the sketch, refer to Figures 3-17 and 3-18.
d. Save the sketch and close the file.

Starting a New File in NX and Invoking the Sketch Environment

1. Start a new file with the name *c03tut2.prt* using the **Model** template and specify its location as *C:\NX\c03*.

   If required turn on the display of WCS by choosing the **Display WCS** tool from the **Tools > Utilities > More > WCS** in the **Ribbon**.

2. Invoke the Sketch environment using the XC-ZC plane as the sketching plane.

Drawing the Sketch

It is recommended that you create the first sketch member of exact measurement by entering the value in the edit box displayed. After creating the first sketch member, you can create the other sketch members by taking the first entity as the reference. After creating the entire sketch, you can modify the values by using the dimensioning tools.

1. Choose the **Profile** tool from the **Direct Sketch** group; you are prompted to specify the first point of the line. Specify the start point of the line at the origin. Next, move the cursor horizontally toward the right and enter **60** in the **Length** edit box and **0** in the **Angle** edit box. Next, press ENTER.

2. Follow the sequence given in Figure 3-16 for drawing the sketch. Draw the rest of the entities of the sketch. For better understanding, the sketch has been numbered and temporary dimensions have been hidden.

Adding Geometric Constraints to the Sketch

After completing the sketch, you need to apply constraints to it.

1. Choose the **Geometric Constraints** tool from the **More** gallery of the **Direct Sketch** group of the **Home** tab; the **Geometric Constraints** dialog box is displayed. Choose the **Vertical** button from this dialog box and select line 8 to apply the vertical dimension, if it is not applied automatically while the sketch is drawn.

2. Similarly, apply the horizontal constraint to lines 1, 3, and 7, if this constraint is not applied automatically.

3. Apply the vertical constraint to line 6, if it is not applied automatically.

4. After applying the constraints, choose the **Display Sketch Constraints** tool from the **Sketch Constraints** gallery of the **More** gallery in the **Direct Sketch** group, if it is not already chosen. The resulting sketch is shown in Figure 3-17.

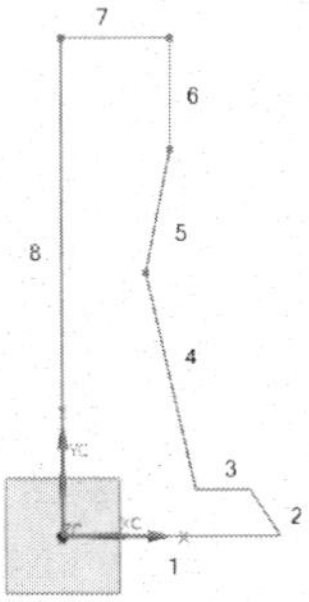

***Figure 3-16** The sequence for drawing the profile*

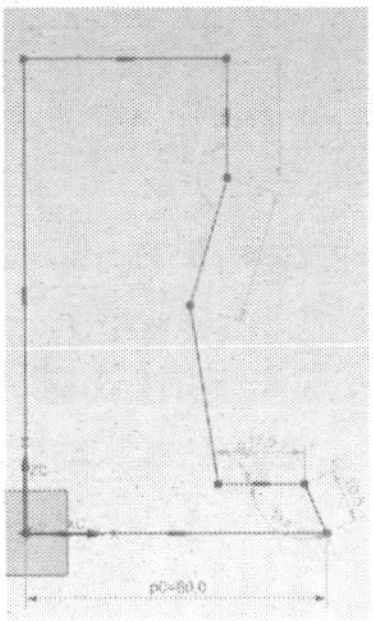

***Figure 3-17** The resulting sketch displayed after adding the constraints*

Modifying Dimensions of the Sketch

Next, you need to modify the dimensions of the sketch.

1. Double-click on the dimension of line 6; a dynamic edit box is displayed. Modify the value of the dimension to **40** and press ENTER. Next, drag and place the dimension on the right of the sketch, refer to Figure 3-18.

   Next, you need to add some dimensions which are not added automatically.

2. Choose the **Rapid Dimension** tool from the **Direct Sketch** group; the **Rapid Dimension** dialog box is displayed.

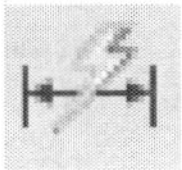

3. Select line 8 and then place the dimension on the left of thesketch; an edit box with default value is displayed. Enter **150** in this edit box and press the ENTER key; the dimension value is modified, refer to Figure 3-18.

4. Select line 5 and then place the dimension on the right of the sketch. Next, modify the dimension value to **30** and press the ENTER key, refer to Figure 3-18.

5. Select line 7 and place the dimension above the sketch. Next, modify the dimension value to **35** and press ENTER, refer to Figure 3-18.

6. Select line 2 and then line 1; an angular dimension gets attached to the cursor. Move the cursor outside the sketch towards right and click the left mouse button to place the dimension. Next, modify the dimension value to **117** and press ENTER, refer to Figure 3-18.

7. Select lines 3 and 4; an angular dimension is attached to the cursor. Move the cursor inside the sketch and place the dimension. Next, modify the dimension value to **77** and press ENTER, refer to Figure 3-18.

8. Select the lower endpoint of line 4 and then select line 8; the dimension value is attached to the cursor. Place the dimension value below the sketch and then modify this value to **40**. Next, press ENTER.

9. Select line 2 and place the dimension on the right of the line. Next, modify the dimension value to **12** and press ENTER, refer to Figure 3-18. Press the ESC key.

10. Choose the **Fit** tool from the **Operation** group of the **View** tab. The resulting sketch after adding all the dimensions is shown in Figure 3-18.

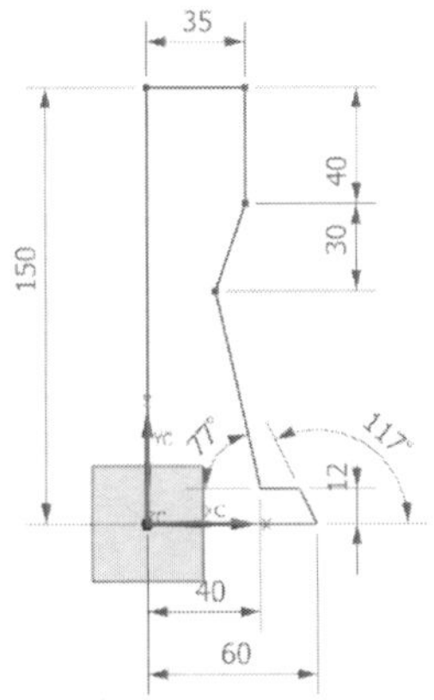

***Figure 3-18** The completed sketch displayed after adding the required constraints and dimensions*

Note

*In Figure 3-18, the display of constraints is turned off to get a better display of dimensions. You can turn off the display of constraints by choosing the **Display Sketch Constraints** tool.*

Saving and Closing the File

1. Choose the **Save** button from the **Quick Access** toolbar to save the sketch. Note that the name and location of the document has already been specified when you started the new file.

2. Exit the sketching environment by choosing the **Finish Sketch** tool from the **Direct Sketch** group of the **Home** tab and choose **Menu > File > Close > All Parts** from the **Top Border Bar** to close the file.

EXERCISES

Exercise 1

Draw the base sketch of the model shown in Figure 3-19. The sketch to be drawn is shown in Figure 3-20. Use the geometric and dimensional constraints to complete this sketch.

(Expected time: 15 min)

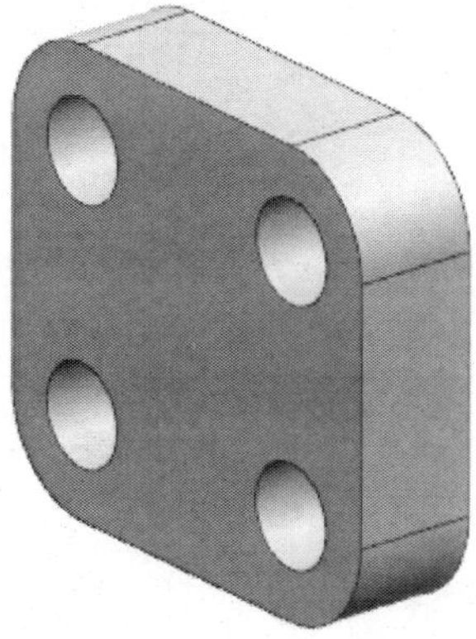

Figure 3-19 *Model for Exercise 1*

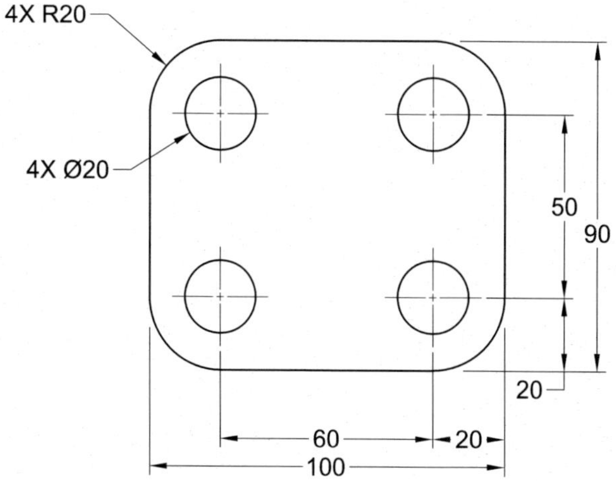

Figure 3-20 *Sketch for Exercise 1*

Exercise 2

Draw the base sketch of the model shown in Figure 3-21. The sketch to be drawn is shown in Figure 3-22. Use the geometric and dimensional constraints to complete this sketch.

(Expected time: 15 min)

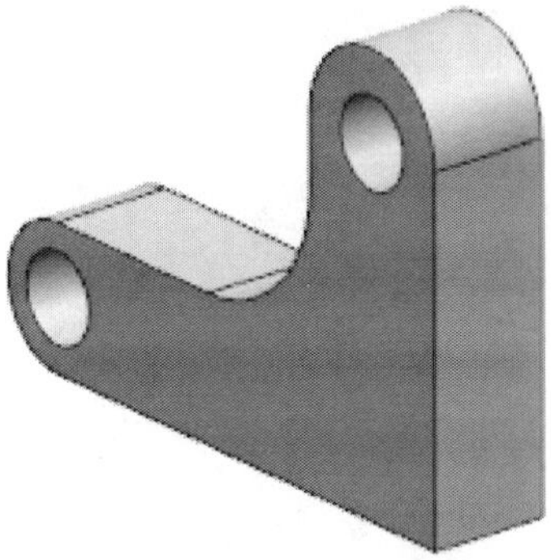

Figure 3-21 *Model for Exercise 2*

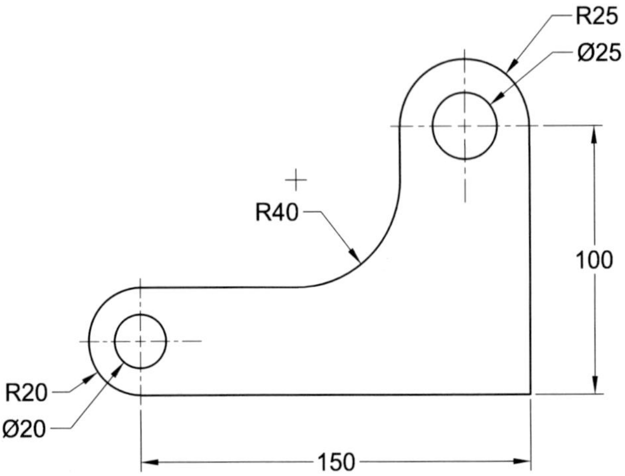

Figure 3-22 *Sketch for Exercise 2*

Chapter 4

Editing, Extruding, and Revolving Sketches

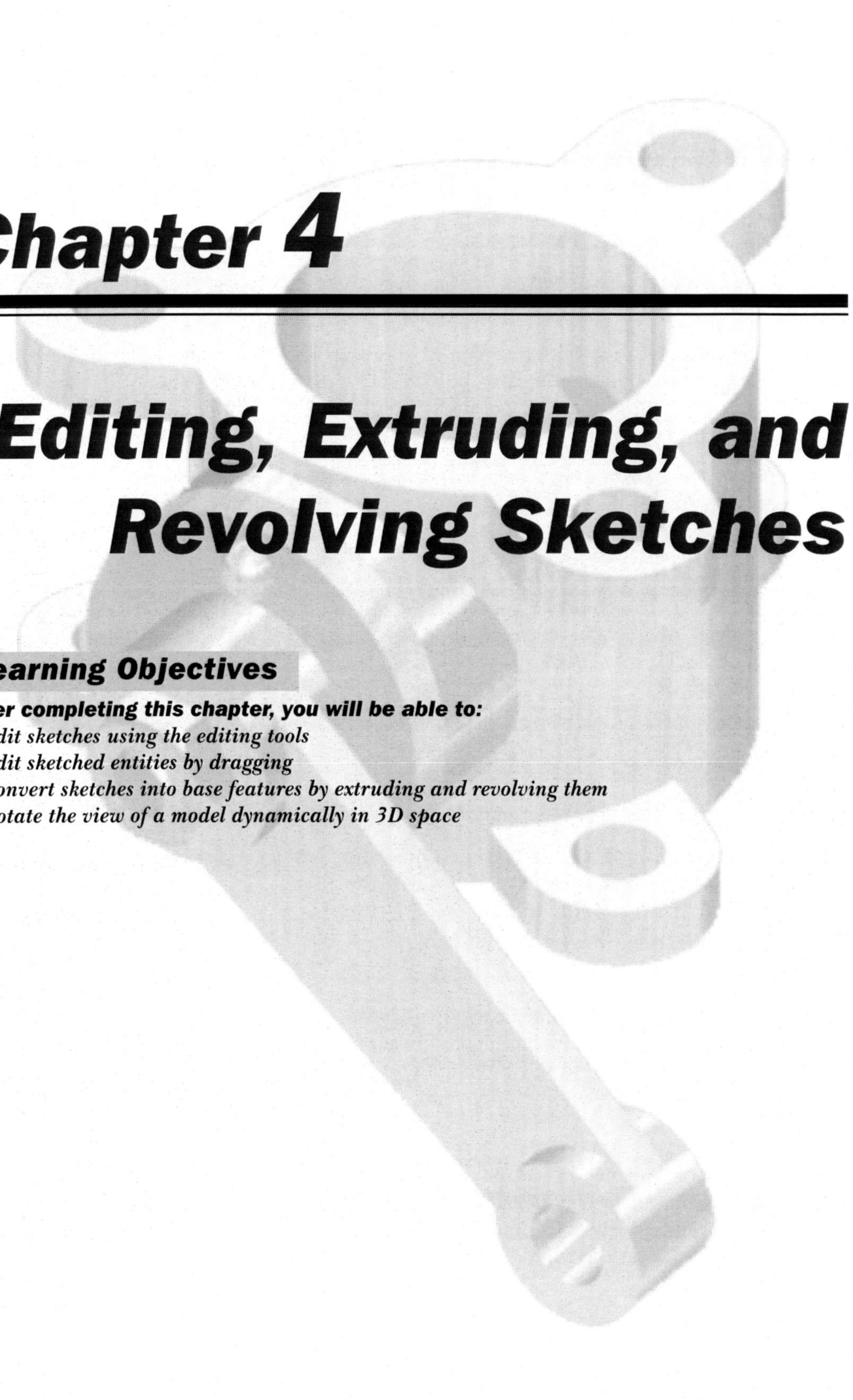

Learning Objectives

After completing this chapter, you will be able to:

- *Edit sketches using the editing tools*
- *Edit sketched entities by dragging*
- *Convert sketches into base features by extruding and revolving them*
- *Rotate the view of a model dynamically in 3D space*

EDITING SKETCHES

Editing is a very important part of sketching in any design or manufacturing program. You need to edit the sketches during various stages of a design. NX provides you with a number of tools that can be used to edit sketched entities. These tools are discussed next.

Trimming Sketched Entities

Ribbon:	Home > Direct Sketch > Sketch Curve Gallery > Quick Trim
Menu:	Edit > Sketch Curve > Quick Trim

The **Quick Trim** tool is used to remove a portion of a sketch by chopping it off. Figure 4-1 shows the sketched entities to be trimmed and Figure 4-2 shows the sketch after performing the trimming operation. Note that when used on an isolated entity, this tool deletes the entity.

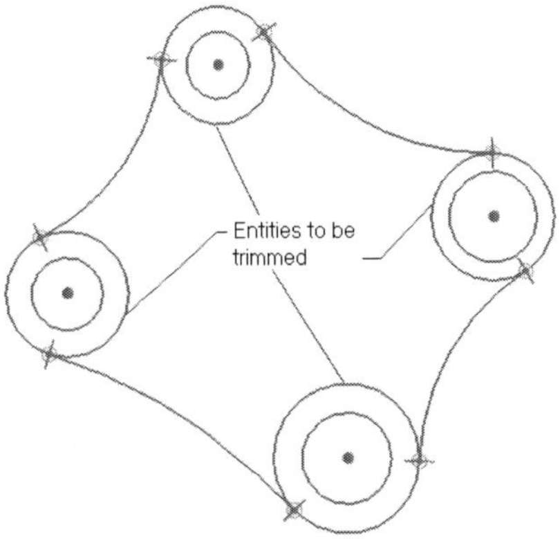

Figure 4-1 Selecting the entities to be trimmed

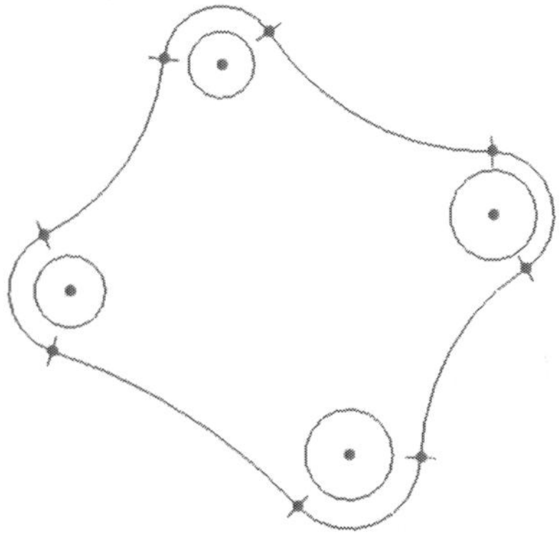

Figure 4-2 Sketch after trimming entities

To trim entities, invoke the **Quick Trim** tool from the **Sketch Curve** gallery of the **Direct Sketch** group of the **Home** tab; the **Quick Trim** dialog box will be displayed, as shown in Figure 4-3. Also, you will be prompted to select a curve to trim.

Move the cursor over the portion to be trimmed; it will be highlighted. Click to trim the highlighted portion. You will be prompted again to select the entity to be trimmed. After trimming all entities, press ESC to exit this tool.

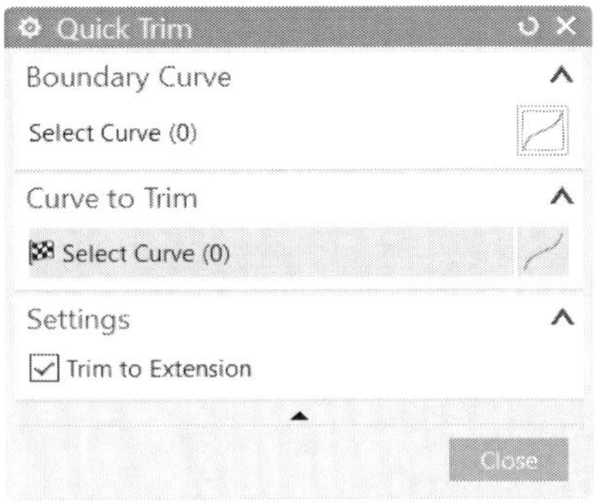

*Figure 4-3 The **Quick Trim** dialog box*

To trim multiple entities, press and hold the left mouse button and drag the cursor over the entities to be trimmed. As you move the cursor over them, all the selected entities will be trimmed. Figure 4-4 shows multiple entities being trimmed by dragging the cursor over them and Figure 4-5 shows the resulting sketch.

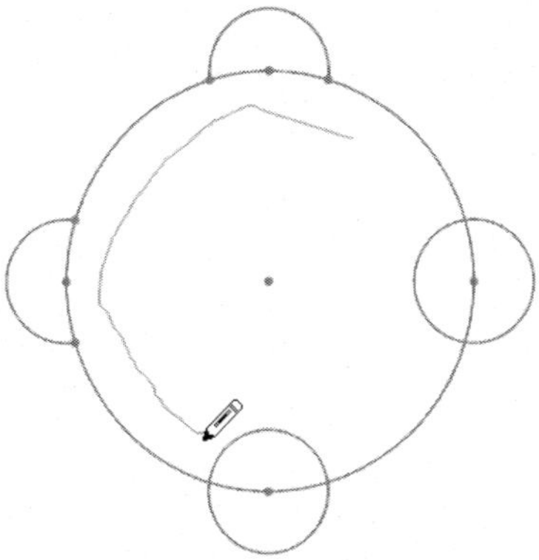

Figure 4-4 Dragging the cursor to trim multiple entities

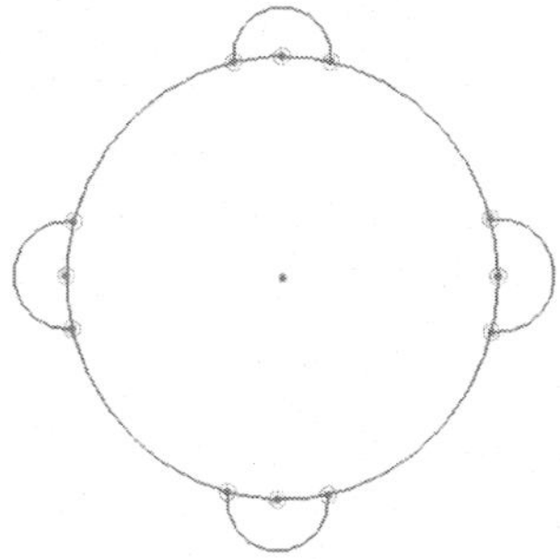

Figure 4-5 The resulting sketch

NX also allows you to select a sketched entity as a knife edge and then trim the other entities using this cutting edge. To trim entities using the cutting edge, choose the **Boundary Curve** button from the **Boundary Curve** rollout in the **Quick Trim** dialog box; you will be prompted to select the boundary curve. Select the curve from the drawing window; the curve becomes a knife edge. Next, choose the **Curve to Trim** button from the **Curve to Trim** rollout in the **Quick Trim** dialog box; you will be prompted to select the curve to be trimmed. Select the curve to be trimmed from the drawing area; the selected curve will be trimmed with respect to the knife edge specified. Similarly, select other curves that intersect the boundary curve; the selected curves will be trimmed. If the **Trim to Extension** check box is selected in the **Settings** rollout and the curve to be trimmed intersects the virtual boundary then the curve will be trimmed on the selected side of the virtual boundary.

Extending Sketched Entities

Ribbon:	Home > Direct Sketch > Sketch Curve Gallery > Quick Extend
Menu:	Edit > Sketch Curve > Quick Extend

The **Quick Extend** tool is used to extend or lengthen an open sketched entity upto the next entity that it intersects. Figure 4-6 shows the sketched entity before extending and Figure 4-7 shows the sketch after extending the entity. Note that this tool will not work on an entity that does not intersect with any existing sketched entity when extended.

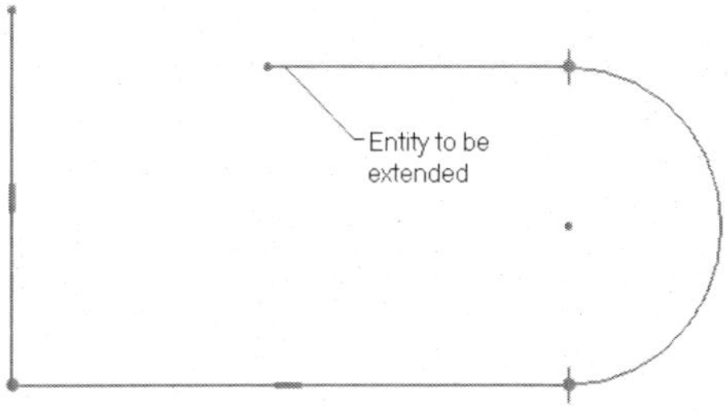

Figure 4-6 Selecting the entity to be extended

Figure 4-7 Sketch after extending the entity

You can also extend multiple entities by pressing and holding the left mouse button and dragging the cursor over them. All the other options of this tool are same as the **Quick Trim** tool.

CREATING BASE FEATURES BY EXTRUDING

Ribbon:	Home > Feature > Design Feature Drop-down > Extrude
Menu:	Insert > Design Feature > Extrude

Extrude is defined as the process of creating a feature from a sketch by adding the material along the direction normal to the sketch or any other specified direction. Figure 4-8 shows the isometric view of a closed sketch and Figure 4-9 shows the extruded feature created using this sketch.

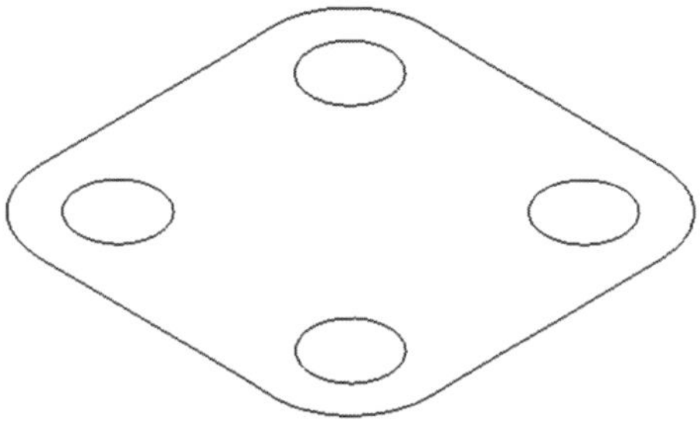

Figure 4-8 *Sketch for the extrude feature*

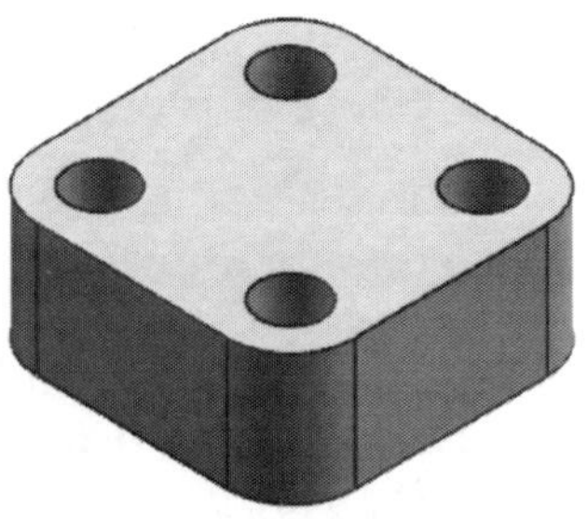

Figure 4-9 *Resulting extruded feature*

When you choose the **Extrude** tool from the **Feature** group of the **Home** tab; the **Extrude** dialog box will be displayed, refer to Figure 4-10. Also, you will be prompted to select the planar face to sketch or the section geometry to be extruded. If you select the sketch at this stage, the preview of the extruded feature created using the default values will be displayed on the screen. If you select the sketch plane, the Sketch in Task Environment will be invoked. Draw the sketch and exit the Sketch in Task Environment; the preview of the extruded feature will be displayed in the Modeling environment. You can also draw the sketch in Sketch in Task Environment using the **Sketch Section** button available at the left of the **Curve** button in the **Section** rollout.

CREATING SOLID REVOLVED BODIES

Ribbon:	Home > Feature > Design Feature Drop-down > Revolve
Menu:	Insert > Design Feature > Revolve

The **Revolve** tool is used to create a solid body by revolving a sketch around the revolution axis which could be a sketched line or an edge of an existing feature. Figure 4-11 shows a sketch for creating a revolved feature and Figure 4-12 shows the isometric view of the resulting feature revolved through an angle of 270 degrees. To convert a sketch into a revolved body, you need to invoke the **Revolve** tool. This tool works in the following three steps:

Step 1: Select the sketch to be revolved
Step 2: Select the revolution axis
Step 3: Specify the revolution parameters

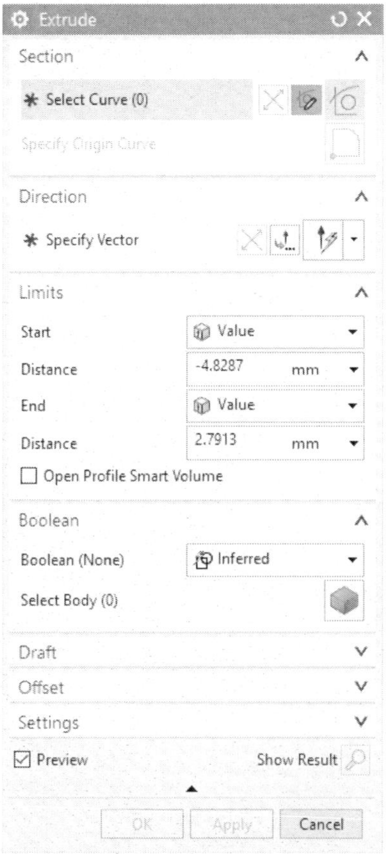

***Figure 4-10** The **Extrude** dialog box*

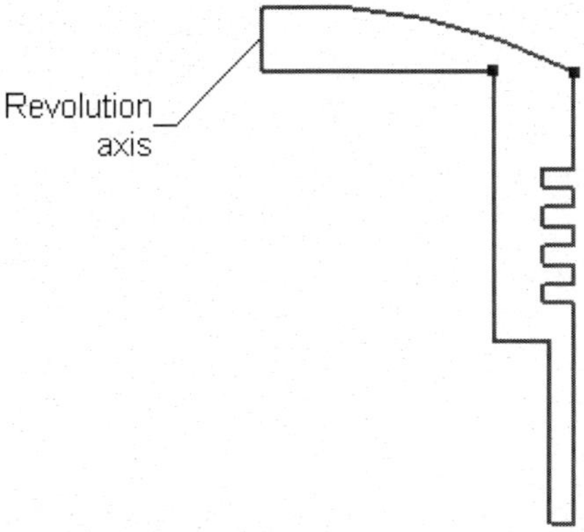

***Figure 4-11** Sketch for creating the revolved feature with the revolution axis*

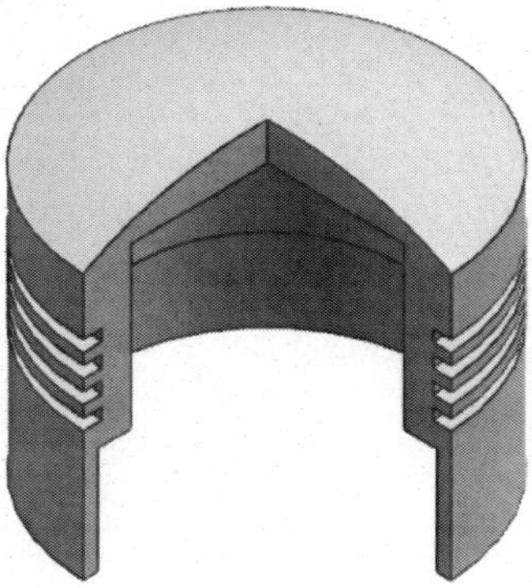

***Figure 4-12** Isometric view of the resulting feature revolved through an angle of 270 degrees*

To create a solid body, choose the **Revolve** tool from the **Design Feature** Drop-down of the **Feature** group; the **Revolve** dialog box will be displayed. Figure 4-13 shows the expanded view of the **Revolve** dialog box. The options in this dialog box are explained next.

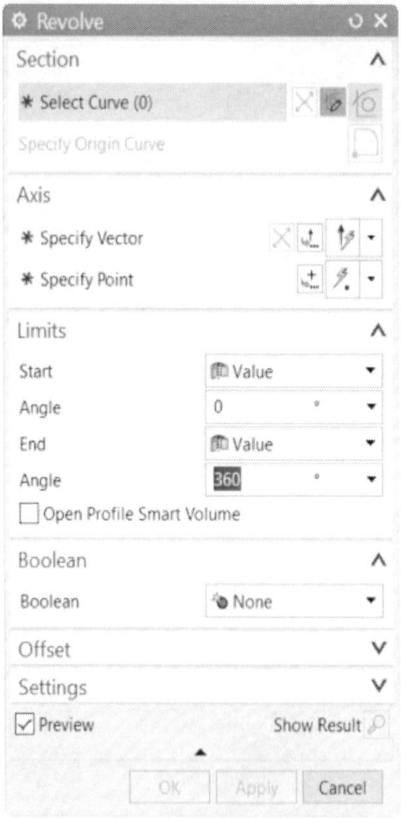

***Figure 4-13** The **Revolve** dialog box*

TUTORIALS

Tutorial 1

In this tutorial, you will create the model shown in Figure 4-14. The dimensions of the model are shown in Figure 4-15. The depth of extrusion of the model is 45.

(Expected time: 30 min)

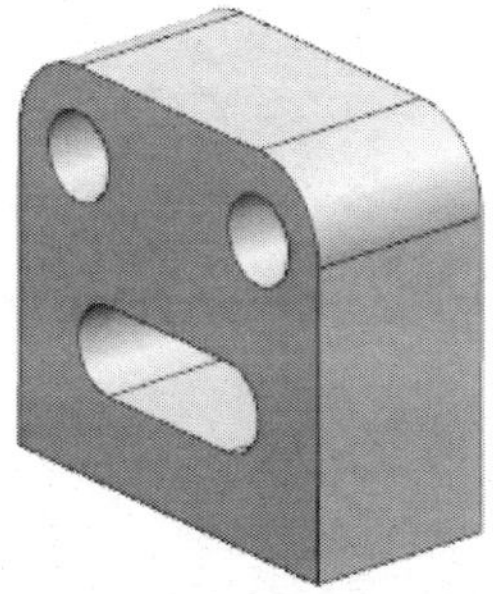

Figure 4-14 Model for Tutorial 1

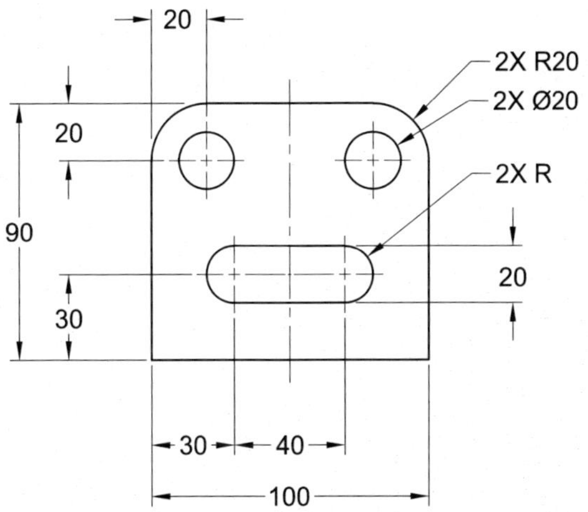

Figure 4-15 Dimensions of the model

The following steps are required to complete this tutorial:

a. Start a new part file using the **Model** template and then draw the profile of the outer loop. Next, add the required constraints, refer to Figure 4-16.
b. Draw the inner loops and add the required dimensions to them, refer to Figure 4-17.
c. Exit the sketching environment and invoke the **Extrude** tool. Define the depth of extrusion to create the final model.
d. Rotate the view in 3D space.
e. Save the file and close it.

Drawing the Sketch of the Model

The sketch of this model will be created on the XC-ZC plane. As mentioned earlier, when you extrude the sketch with nested closed loops, the inner loops are automatically subtracted from the outer sketch.

1. Start a new file with the name *c04tut1.prt* using the **Model** template and specify its location as *C:\NX\c04*.

2. Invoke the sketching environment using the XC-ZC plane as the sketching plane.

3. Draw the sketch of the outer loop and add the required geometric and dimensional constraints to it, as shown in Figure 4-16.

4. Draw the inner loops and then add the required geometric and dimensional constraints to them, as shown in Figure 4-17.

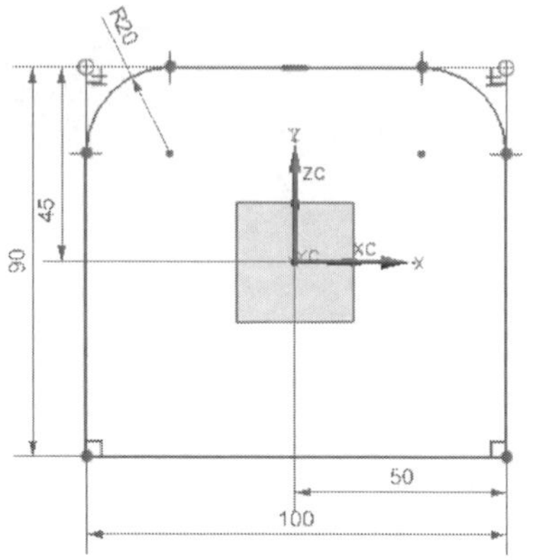

Figure 4-16 *Dimensioned sketch of the outer loop*

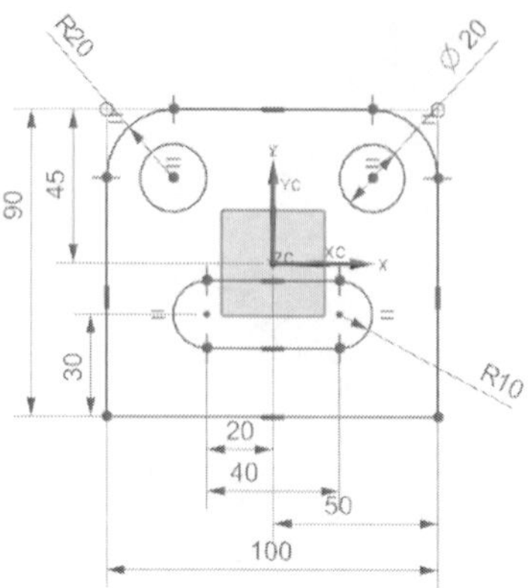

Figure 4-17 *Final sketch of the base feature*

Converting the Sketch into Base Feature

Next, you need to convert the sketch into base feature by using the **Extrude** tool.

1. Choose the **Finish Sketch** tool from the **Direct Sketch** group to exit the sketching environment.

2. Right-click in the drawing area; a shortcut menu is displayed. Choose the **Fit** option from the shortcut menu to fit the sketch in the screen.

3. Invoke the **Extrude** tool from the **Feature** group; the **Extrude** dialog box is displayed and you are prompted to select the planar face to sketch or select the section geometry.

4. Select the sketch from the drawing window; the preview of the extruded feature is displayed. Also, the **End** dynamic input box is displayed in the drawing window.

5. Enter **45** in the **End** dynamic input box and press the ENTER key; the preview is modified accordingly.

6. Choose the **OK** button in the **Extrude** dialog box; the extrude feature is created and displayed in the drawing window.

7. Press the CTRL+B keys; the **Class Selection** dialog box is displayed. Select the sketch of the extruded feature to hide it. Next, choose the **OK** button from the dialog box. Figure 4-18 shows the extruded feature after hiding its sketch.

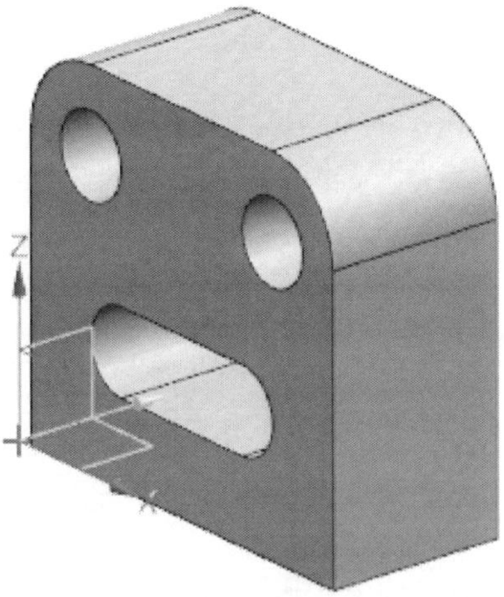

Figure 4-18 *Extruded model for Tutorial 1*

Rotating the View of the Model

Next, you need to rotate the view of the model so that you can maneuver and view it from different directions.

1. Choose the **Rotate** tool from the **Operation** group of the **View** tab in the **Ribbon**; the cursor changes into a rotate view cursor.

2. Press and hold the left mouse button and drag the cursor in the drawing window to rotate the view of the model.

3. Exit the **Rotate** tool by pressing the ESC key.

4. Choose the **Trimetric** button from the **Orient View Drop-down** in the **View** group in the **Top Border Bar** to restore the Isometric view.

5. Right-click in the drawing window and choose the **Fit** option from the shortcut menu; the model fits in the drawing window.

Saving and Closing the File

1. Choose **Menu > File > Close > Save and Close** from the **Top Border Bar** to save and close the file.

Tutorial 2

In this tutorial, you will create the model of a revolved body shown in Figure 4-19. You can create this model using the sketch shown in Figure 4-20. **(Expected time: 30 min)**

Figure 4-19 *Model for Tutorial 3*

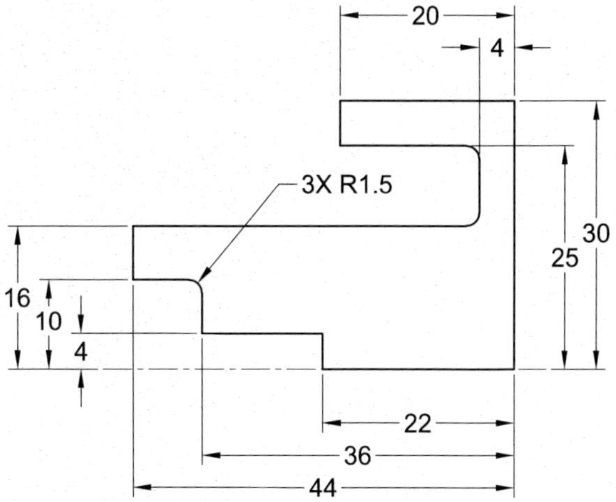

Figure 4-20 *Dimensions of the model*

The following steps are required to complete this tutorial:

a. Start a new part file using the **Model** template and draw the sketch of the revolved model. Next, add the required geometric and dimensional constraints.
b. Exit the sketching environment and invoke the **Revolve** tool. Select the sketch to be revolved and then the direction of the axis of revolution.
c. Define the other revolution parameters to create the final model.
d. Rotate the view in 3D space.
e. Save and close the file.

Drawing the Sketch of the Model

The sketch of the revolved model will be created on the YC-ZC plane.

1. Start a new file with the name *c04tut2.prt* using the **Model** template and specify its location as *C:\NX\c04*.

2. Invoke the sketching environment by using the YC-ZC plane as the sketching plane.

3. Draw the sketch of the revolved feature and add the required geometric and dimensional constraints to it, as shown in Figure 4-21.

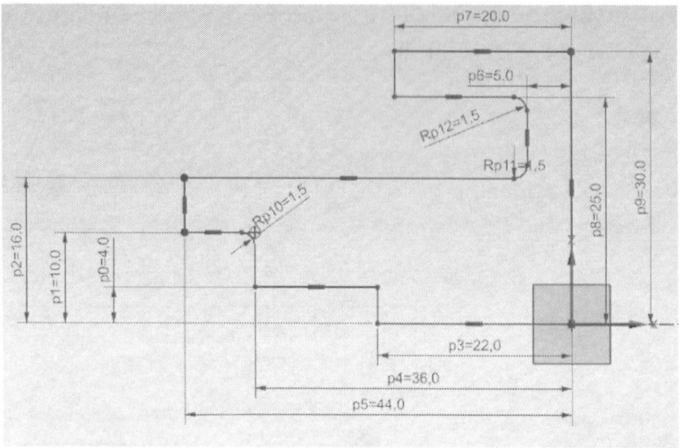

Figure 4-21 *Dimensioned sketch of the revolved feature*

Converting the Sketch into the Revolved Feature

Next, you need to convert the sketch into a revolved feature by using the **Revolve** tool.

1. Choose the **Finish Sketch** tool from the **Direct Sketch** group to exit the sketching environment.

2. Right-click in the drawing window; a shortcut menu is displayed. Choose the **Fit** option from the shortcut menu to fit the sketch in the screen.

3. Invoke the **Revolve** tool from the **Feature** group; the **Revolve** dialog box is displayed and you are prompted to select the planar face to sketch or select the section geometry.

4. Select the sketch drawn and choose the **Inferred Vector** button from the **Axis** rollout of the dialog box; you are prompted to select the objects to infer the vector.

5. Select the bottom horizontal line as the axis of revolution; the preview of the revolved body is displayed.

6. Accept the default options in the **Revolve** dialog box and choose the **OK** button; the revolved feature is created. Make sure that the angle of revolution of the model is 360 degrees.

7. Choose the **Show and Hide** tool from the **Visibility** group of the **View** tab; the **Show and Hide** dialog box is displayed. Click on the minus sign (-) in the **Datums** and **Sketches** rows to hide the sketch, axes, and sketching plane, and then close the dialog box. Next, fit the model to the screen. The revolved model after hiding the sketch, the axes, and the sketching plane is shown in Figure 4-22.

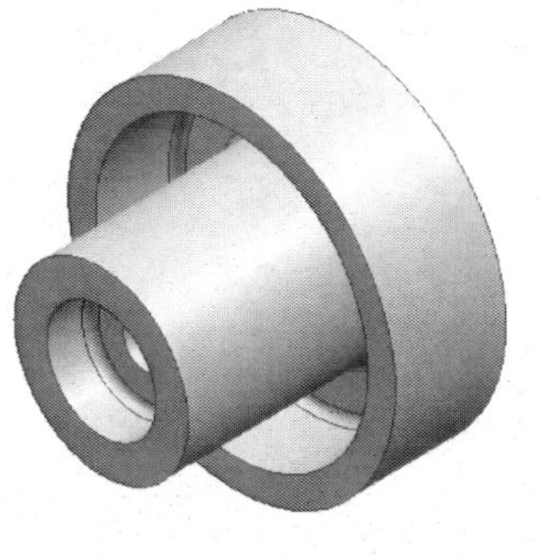

Figure 4-22 *Revolved model of Tutorial 2*

Rotating the View of the Model

Next, you need to rotate the model to maneuver and view it from different directions.

1. Choose the **Rotate** tool from the **Operation** group of the **View** tab of the **Ribbon**; the cursor changes into a rotate view cursor.

2. Press and hold the left mouse button and drag the cursor in the drawing window to rotate the view of the model. Next, exit from the **Rotate** tool by pressing the ESC key.

3. Choose the **Trimetric** button from the **Orient View Drop-down** of the **View** group in the **Top Border Bar** to restore the Isometric view.

Saving and Closing the File

1. Choose **Menu > File > Close > Save and Close** from the **Top Border Bar** to save and close the file.

EXERCISES

Exercise 1

Open the sketch drawn in Tutorial 2 of Chapter 3 and convert it into a fully revolved body. After creating the model, use the **Rotate** tool to rotate its view. Save the file with a different name in the folder of this chapter. **(Expected time: 15 min)**

Exercise 2

Create the model shown in Figure 4-23. The dimensions of the model are shown in Figure 4-24. The depth of extrusion is 30. After creating the model, use the **Rotate** tool to rotate its view. Before closing the file, restore the Isometric view of the model.

(Expected time: 30 min)

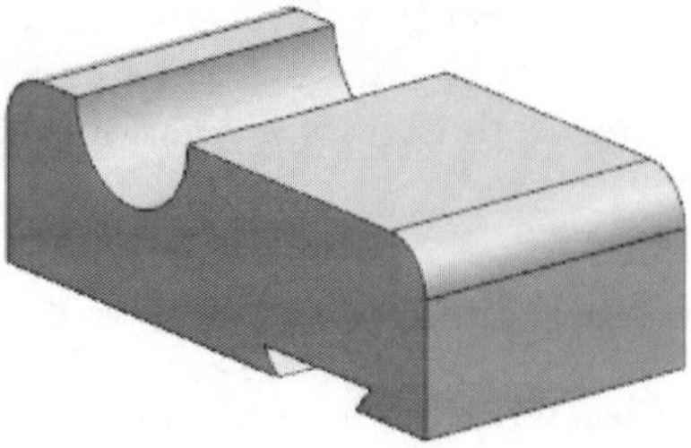

***Figure 4-23** Model for Exercise 2*

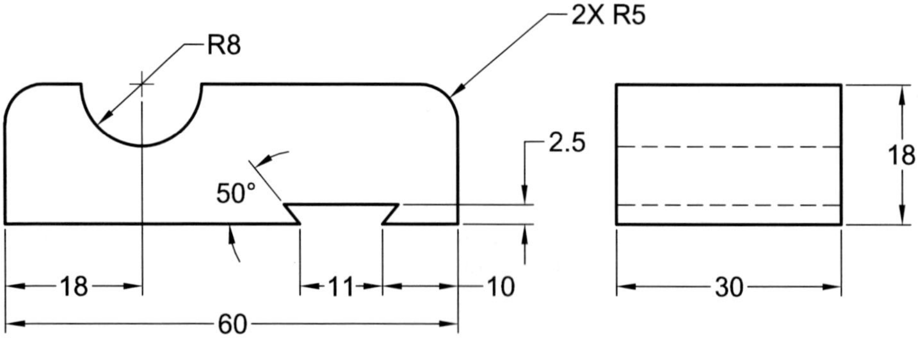

***Figure 4-24** Dimensions of the model*

Chapter 5

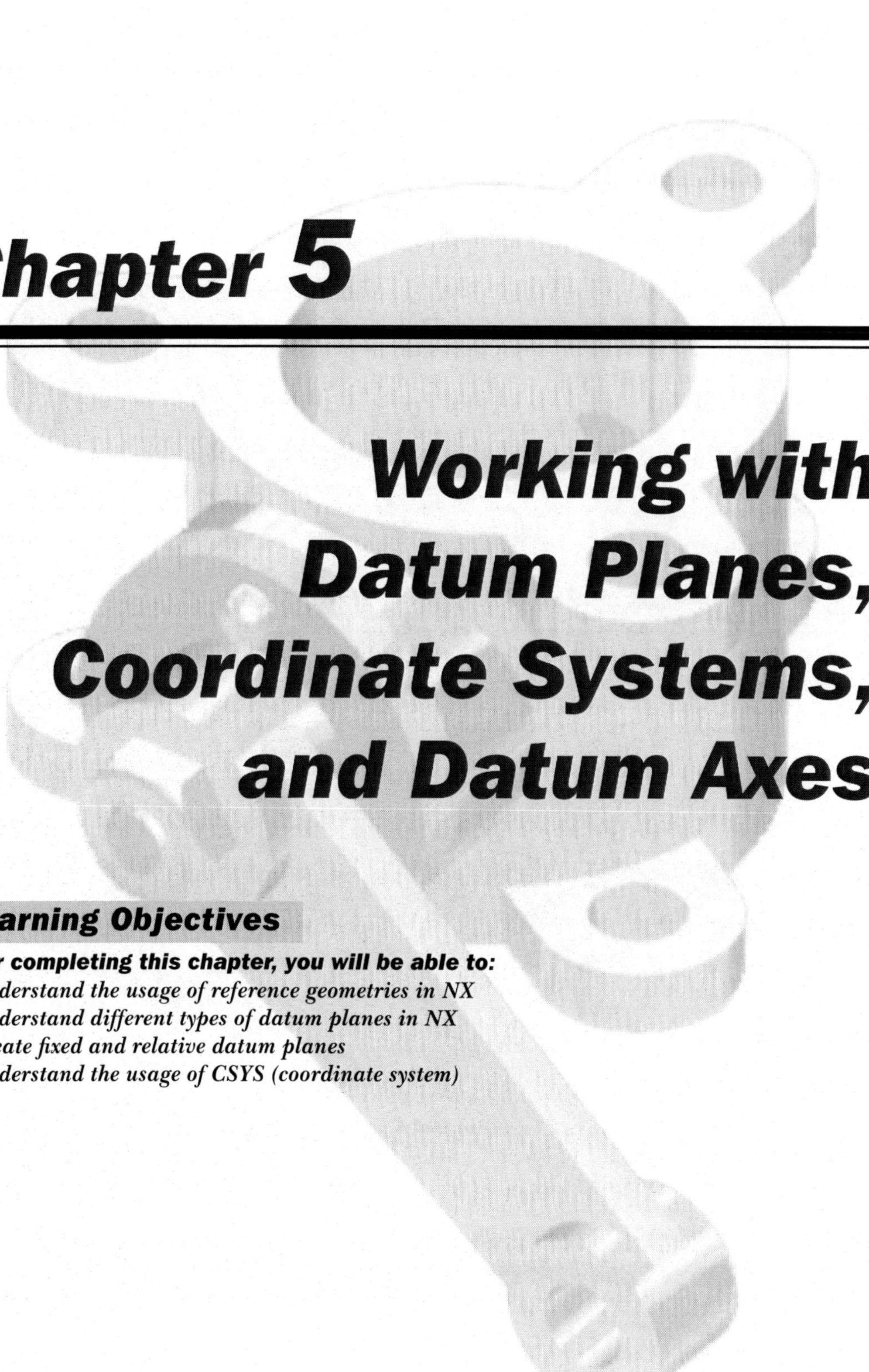

Working with Datum Planes, Coordinate Systems, and Datum Axes

Learning Objectives

After completing this chapter, you will be able to:

- *Understand the usage of reference geometries in NX*
- *Understand different types of datum planes in NX*
- *Create fixed and relative datum planes*
- *Understand the usage of CSYS (coordinate system)*

ADDITIONAL SKETCHING AND REFERENCE PLANES

In the previous chapters, you learned to create basic models which had features placed on one of the three datum planes of the Datum Coordinate System (YC-ZC, XC-ZC, or XC-YC). All the models were created by selecting one of the three datum planes of the Datum Coordinate System. Most real world models consist of multiple sketched features, reference geometries, and placed features. In NX, features can be added to base features by using the boolean operations such as subtract, unite, and intersect. These boolean operations are available in all tools. When you invoke the sketching environment by choosing the **Sketch** tool from the **Direct Sketch** group of the **Home** tab, the **Create Sketch** dialog box is displayed and you are prompted to select a sketching plane. You can select any one of the three datum planes of the Datum Coordinate System as the sketching plane. You can also create a new datum plane by using the **Datum Plane** dialog box. On the basis of design requirement, you can select any plane as the sketching plane for the base feature. Also, you can create additional planes by taking reference of existing planes, faces, surfaces, sketches, or a combination of these objects. Figure 5-1 shows a model with multiple features.

The base feature for this model is shown in Figure 5-2. The sketch for the base feature is drawn on the XC-YC datum plane of the Datum Coordinate System. As mentioned earlier, after creating the base feature, you need to create the other sketched features, placed features, and reference features, refer to Figure 5-3. The extrude features shown in Figure 5-3 require additional sketching planes on which the sketch for the other features will be created.

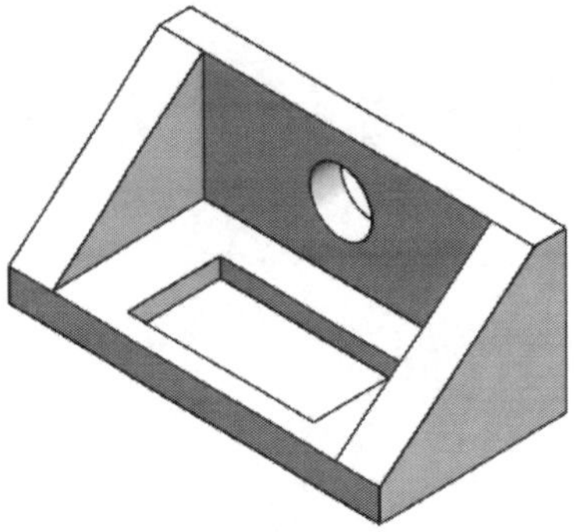

Figure 5-1 *Model with multiple features*

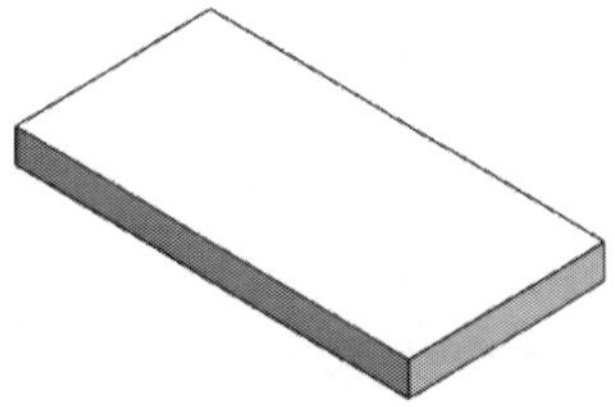

Figure 5-2 *The base feature for the model*

It is evident from Figure 5-3 that the additional features created on the base feature do not lie on the same sketching plane. They are created by defining additional sketching planes. Also, appropriate boolean operations are selected at the time of creating these features.

As mentioned earlier, if you invoke a new part file by using the **Model** template, the Datum Coordinate System containing three datum planes, three datum axes, and a point will be displayed in the drawing window. You can select any of these datum planes as the sketching plane or you can create new datum planes by using the **Datum Plane** dialog box.

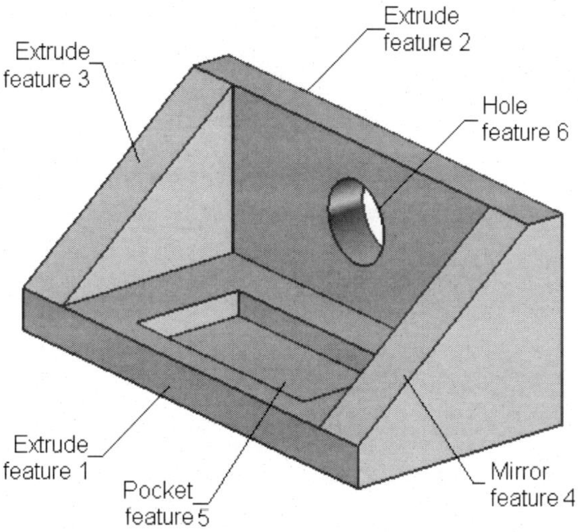

Figure 5-3 *Various features in the model*

TYPES OF DATUM PLANES

There are three types of datum planes in NX: datum planes of the Datum Coordinate System, fixed datum planes, and relative datum planes. Fixed datum planes are the ones that pass through the origin. Relative datum planes are created in addition to fixed datum planes by taking the reference of objects such as curves, sketches, edges, faces, surfaces, and points. The methods for creating these datum planes are discussed next.

Creating Three Fixed (Principle) Datum Planes

In NX, the principle datum planes are called fixed datum planes.

Creating Relative Datum Planes

Ribbon:	Home > Feature > Datum/Point Drop-down > Datum Plane
Menu:	Insert > Datum/Point > Datum Plane

As mentioned earlier, relative datum planes are the additional planes used for assisting you in completing a design. You can select objects such as curves, sketches, edges, faces, surfaces, datum planes of the Datum Coordinate System, and points as reference to create relative datum planes. To create relative datum planes, choose the **Datum Plane** tool from the **Datum/Point Drop-down** of the **Feature** group in the **Home** tab; the **Datum Plane** dialog box will be displayed, as shown in Figure 5-4. In NX, you can create relative datum planes using different options. You can access these options from the drop-down list available in the **Type** rollout of the dialog box, as shown in Figure 5-5. Some of these options are discussed next.

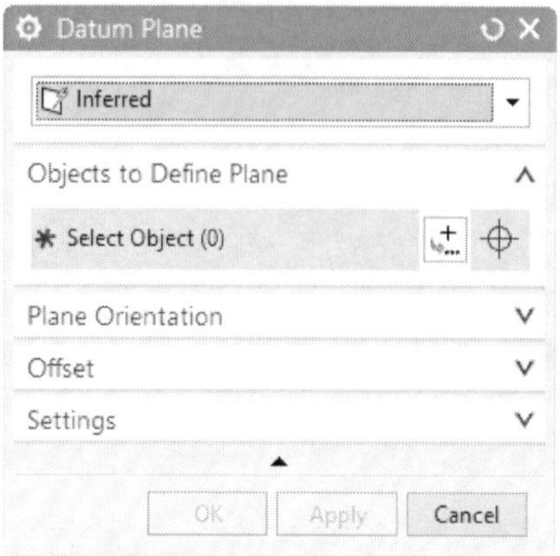

***Figure 5-4** The **Datum Plane** dialog box*

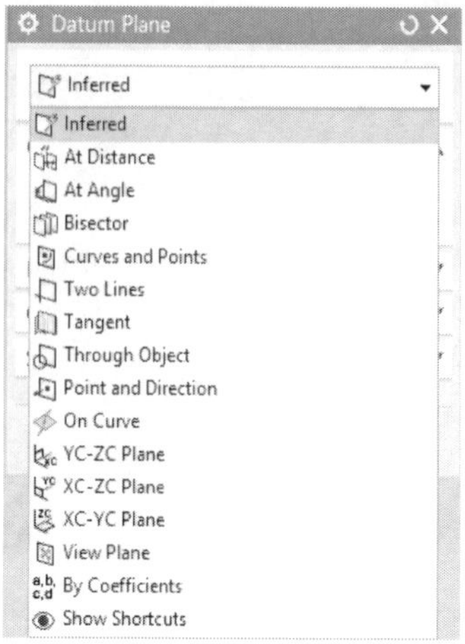

***Figure 5-5** The drop-down list in the **Type** rollout of the **Datum Plane** dialog box*

CREATING DATUM COORDINATE SYSTEMS

Ribbon: Home > Feature > Datum/Point Drop-down > Datum CSYS
Menu: Insert > Datum/Point > Datum CSYS

As discussed earlier, on starting a new file using the **Model** template, you will only have the Datum Coordinate System at the origin. The handles of this coordinate system represent the X, Y, and Z directions in 3D space and form the basis for creating both the fixed and the relative datum planes. You can also create new coordinate systems for the separate features in a model. A part may contain any number of coordinate systems. The coordinate system can be used as a reference for creating features, sketches, and curves. Also, it can be used for assembling the parts in the assembly. The coordinate system is also treated as a feature and is displayed in the **Model History** section of the **Part Navigator**. The coordinate system is always associative with the object members or feature operation to which it is related.

To create the Datum Coordinate System, choose the **Datum CSYS** tool from the **Datum/Point Drop-down** of the **Feature** group in the **Home** tab; the **Datum CSYS** dialog box will be displayed, as shown in Figure 5-6. By default, the **Dynamic** option is selected in the Type drop-down list. As a result, you will be prompted to drag or select a handle for direct entry. There are different methods by which you can create a Datum Coordinate System. The options to use these methods are discussed next.

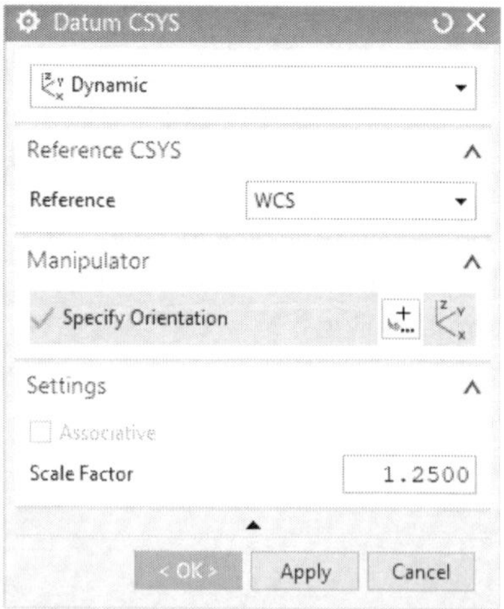

***Figure 5-6** The **Datum CSYS** dialog box*

Dynamic

This option is used to create the Datum Coordinate System by dragging it or entering the X, Y, and Z coordinates values. When you invoke the **Datum CSYS** tool, this option will be selected by default and you will be prompted to drag or select a handle for direct entry. Also, the preview of the Datum Coordinate System will be displayed at the origin in the drawing window, refer to Figure 5-7.

Select the sphere displayed at the center of new Datum Coordinate System, and then press and hold the left mouse button. Next, drag the mouse to position the new Datum Coordinate System. Alternatively, enter the X, Y, and Z coordinates values in the respective edit boxes to position the Datum Coordinate System.

Also, you can move the new Datum Coordinate System in a particular direction (X, Y, or Z). To do so, click on the respective arrow handle; the **Distance** and **Snap** edit boxes will be displayed, as shown in Figure 5-8. Enter the required distance value in the **Distance** edit box and press ENTER; the Datum Coordinate System will move in the specified direction to the specified distance.

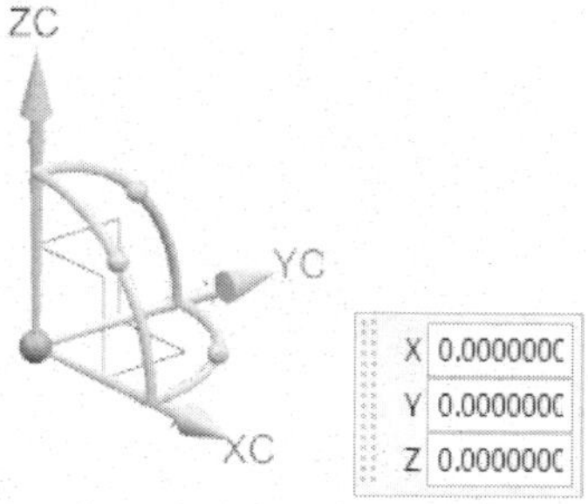

Figure 5-7 *Preview of the datum coordinate system*

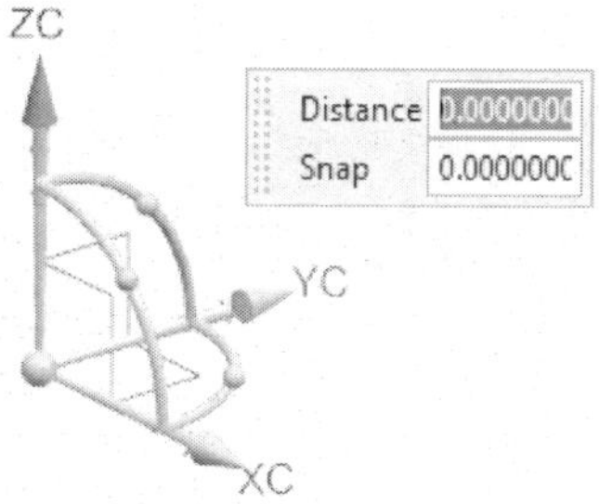

Figure 5-8 *The* ***Distance*** *and* ***Snap*** *edit boxes*

Similarly, you can rotate the new Datum Coordinate System along a particular axis (X, Y, or Z). To do so, click on the respective angular handle; the **Angle** and **Snap** edit boxes will be displayed. Enter the required angle value in the **Angle** edit box and press ENTER; the Datum Coordinate System will rotate along the specified axis as per the specified angle value.

Note

You can also move and rotate the new Datum Coordinate System dynamically. To do so, press and hold the left mouse button on the respective handle of the new Datum Coordinate System and then drag it.

Inferred

You can use the **Inferred** option to create different types of datum coordinate systems. The resulting datum coordinate system will depend on the reference objects selected and the sequence of their selection.

Origin, X-point, Y-point

To create a Datum Coordinate System by using this option, select the **Origin, X-point, Y-point** option from the Type drop-down list; you will be prompted to select the object to infer point. Next, you need to specify the origin point. Select the point on which you need to fix the origin of the Datum Coordinate System. While specifying the origin point, turn on the required snap button in the **Selection Group** to select points easily. Next, you need to specify the points along the X and Y directions. Specify the X and Y points. Note that the X point is used to specify the X-axis direction of the Datum Coordinate System and the Y point is used to define the orientation of the XY plane.

After specifying the origin, X point and Y point, preview of the Datum Coordinate System will be displayed, as shown in Figure 5-9. Choose the **OK** button from the **Datum CSYS** dialog box to accept the Datum Coordinate System. Figure 5-10 shows the resulting Datum Coordinate System.

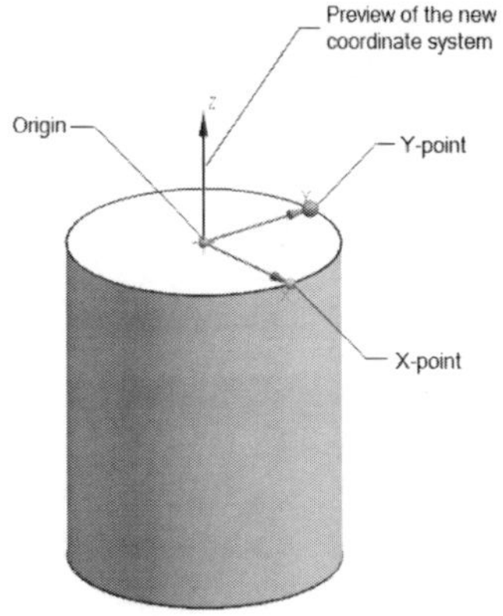

Figure 5-9 *Preview of the CSYS created by using the* ***Origin, X-point, Y-point*** *option*

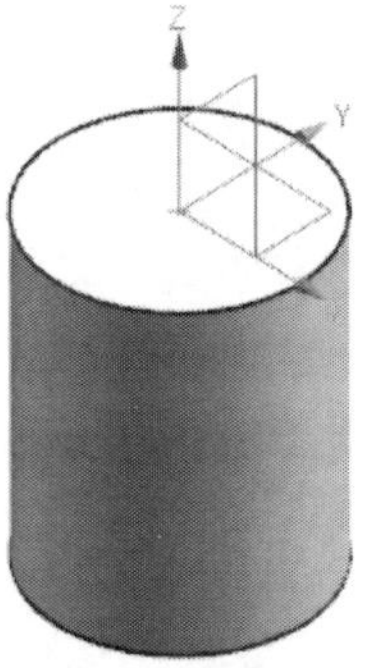

Figure 5-10 *The CSYS created by using the* ***Origin, X-point, Y-point*** *option*

Three Planes

To create a coordinate system by using this option, you need to select three reference objects that are mutually perpendicular to each other. You can only select faces or planes as reference objects.

Select the **Three Planes** option from the Type drop-down list; you will be prompted to select a planar object. Select the type of reference object from the **Type Filter** drop-down list in the **Selection Group**. This allows you to customize the process of selecting a particular type of reference object. Next, select three reference objects. If the selected reference objects are not mutually perpendicular, then the **Alerts** message will be displayed, informing that the parallel planes are selected. After selecting the three reference objects, the preview of the Datum Coordinate System will be displayed, refer to Figure 5-11. Choose the **OK** button from the **Datum CSYS** dialog box to accept the Datum Coordinate System. The resulting Datum Coordinate System is shown in Figure 5-12.

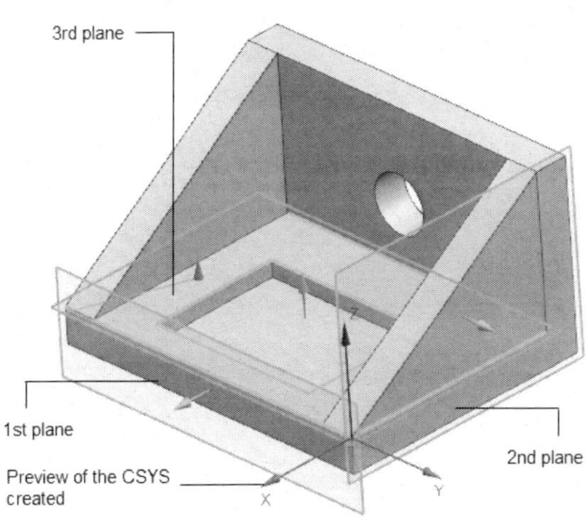

*Figure 5-11 The preview of the CSYS created by using the **Three Planes** option*

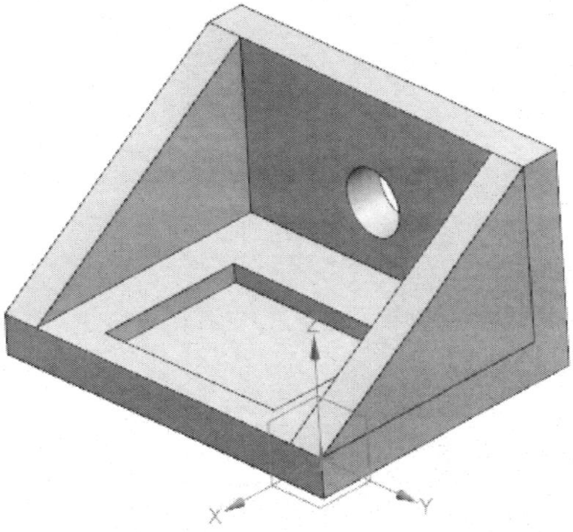

*Figure 5-12 The resulting CSYS created by using the **Three Planes** option*

TUTORIALS

Tutorial 1

In this tutorial, you will create the model shown in Figure 5-13. The dimensions of the model are shown in Figure 5-14. **(Expected time: 30 min)**

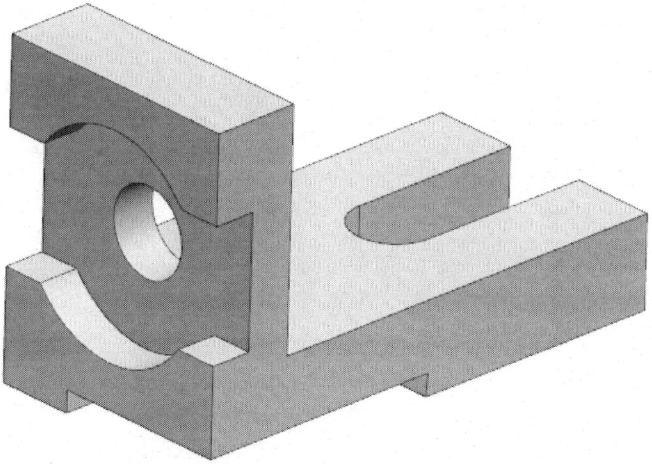

Figure 5-13 Model for Tutorial 1

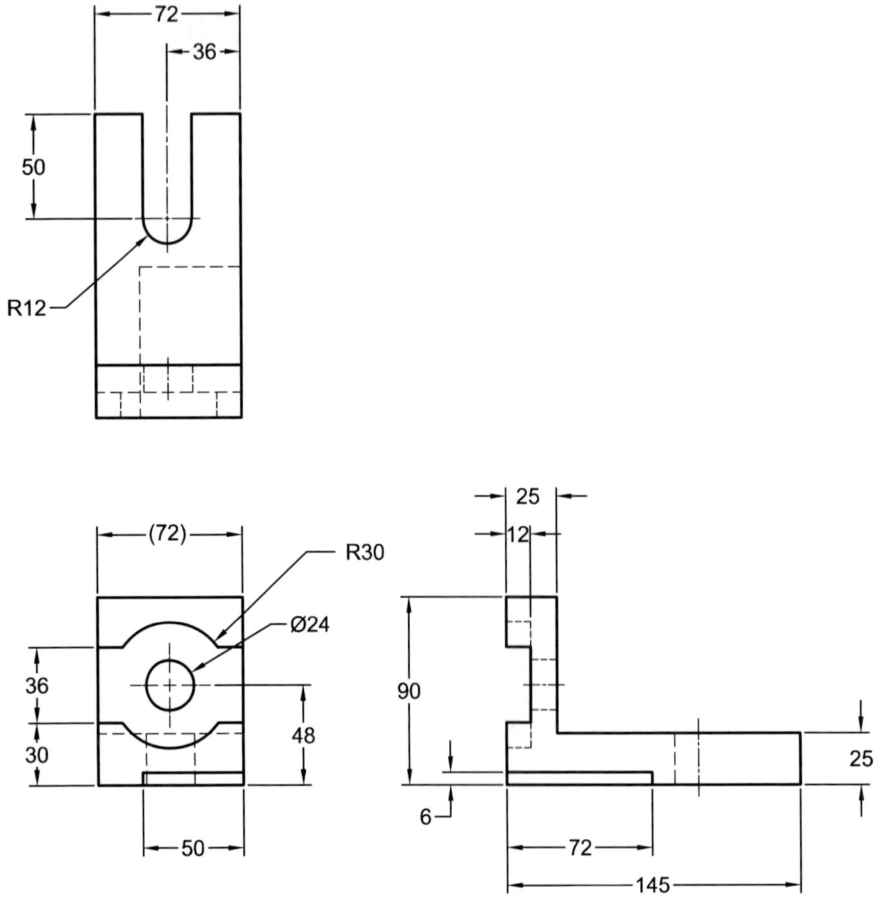

Figure 5-14 Dimensions of the model for Tutorial 1

The following steps are required to complete this tutorial:

a. Create the base feature of the model on the YC-ZC plane.
b. Create the second feature on the front face of the base feature by using the **Subtract** operation.
c. Create the third feature on the upper horizontal face of the base feature by using the **Subtract** operation.
d. Create the fourth and fifth features.

Creating the Base Feature

First, you need to create the base feature of the model.

1. Invoke the **New** dialog box and specify the location of the document as *C:\NX\c05* and name it as *c05tut1.prt*. Next, choose the **OK** button from it.

2. Invoke the sketching environment by using the YC-ZC plane and draw an open sketch, as shown in Figure 5-15.

3. Exit the sketching environment and invoke the **Extrude** dialog box. Select the sketch and then select the **Symmetric Value** option from the **Start** drop-down list in the **Limits** rollout; the **Start** drop-down list is converted into the **End** drop-down list. Enter **36** in the **Distance** edit box located below the **End** drop-down list and then press ENTER.

   Note that by default, only few rollouts are expanded in the **Extrude** dialog box.

4. Expand the **Offset** rollout and select the **Two-Sided** option from the **Offset** drop-down list; the **Start** and **End** edit boxes are displayed.

5 Enter **25** in the **Start** edit box. Also, make sure that the value entered in the **End** edit box is **0**.

6. Choose the **OK** button from the dialog box to create the base feature. Turn off the display of the sketch. The base feature of the model is shown in Figure 5-16.

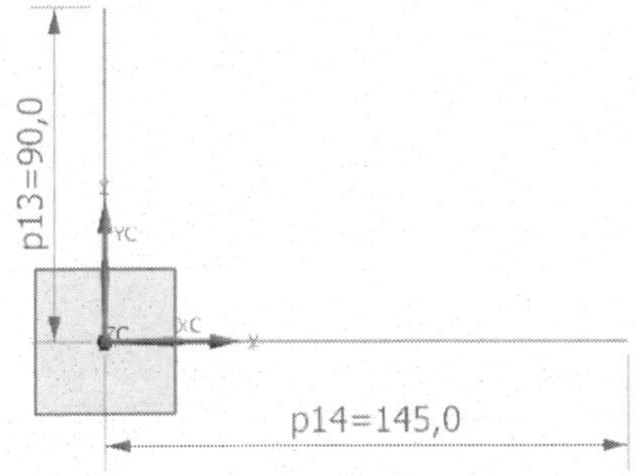

Figure 5-15 *Open sketch for the base feature*

Figure 5-16 *Base feature of the model*

Creating the Second Feature

The second feature is an extruded feature that can be created by using the **Subtract** option. The sketch of this feature is drawn on the front face of the base feature.

1. Invoke the sketching environment by selecting the right face of the base feature as the sketching plane. Make sure that the X-axis of the new sketching plane points toward the edge which measures 145 mm.

2. Draw the rectangular sketch of the cut feature. Make the bottom line and the left line of the rectangle collinear with the edges of the base feature. Add the required dimensions to the sketch, refer to Figure 5-17.

3. Exit the sketching environment and then invoke the **Extrude** dialog box. Now, select the sketch; the preview of the feature is displayed in the graphics window.

4. Select the **Subtract** option from the **Boolean** drop-down list and reverse the direction of extrusion by choosing the **Reverse Direction** button from the **Direction** rollout of the dialog box, if required. Enter **50** in the **Distance** edit box which is available below the **End** drop-down list of the **Limits** rollout and then choose the **OK** button from the **Extrude** dialog box.

5. Press CTRL+B; the **Class Selection** dialog box is displayed. Select the sketch of the second feature from the graphics window and choose the **OK** button from this dialog box; the selected sketch is hidden. The model with the cut feature created is shown in Figure 5-18.

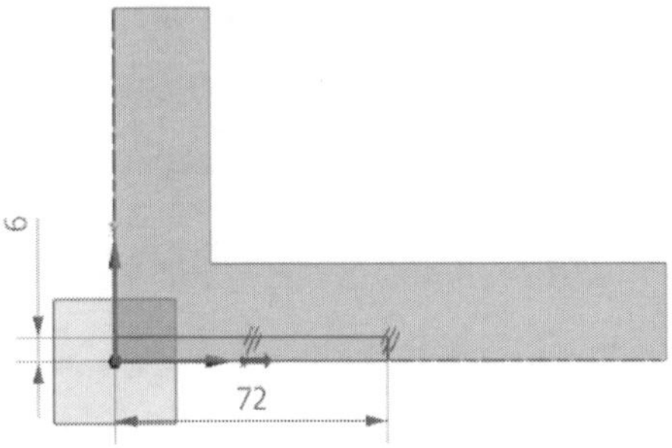

Figure 5-17 *Sketch created for the second feature*

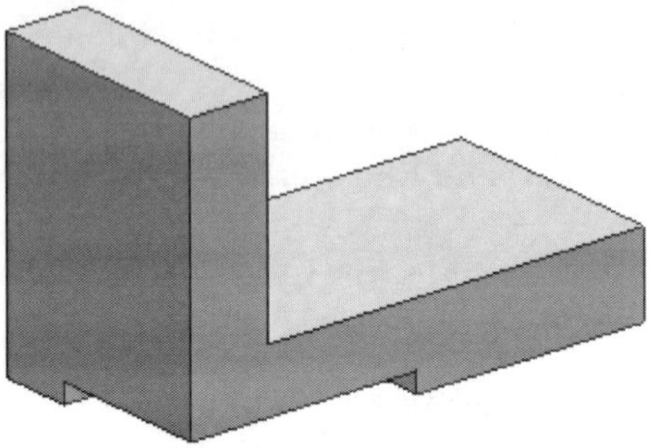

Figure 5-18 *Model with the cut feature*

Creating the Third Feature

The third feature is also an extruded feature which will be created by using the **Subtract** option. The sketch of this feature will be drawn on the top lower face of the base feature.

1. Invoke the sketching environment by selecting the top upper face of the base feature as the sketching plane. Make sure that the X-axis of the new sketching plane points toward the edge which measures 145 mm.

2. Draw the sketch of the feature and add the required constraints and dimensions to it, as shown in Figure 5-19.

3. Exit the sketching environment and then invoke the **Extrude** dialog box. Select the sketch; the preview of the feature is displayed in the drawing window.

4. Select the **Subtract** option from the **Boolean** drop-down list if not already selected and reverse the direction of extrusion.

5. Select the **Through All** option from the **End** drop-down list and choose the **OK** button from the dialog box.

6. Hide the sketch. The model with the cut feature is shown in Figure 5-20.

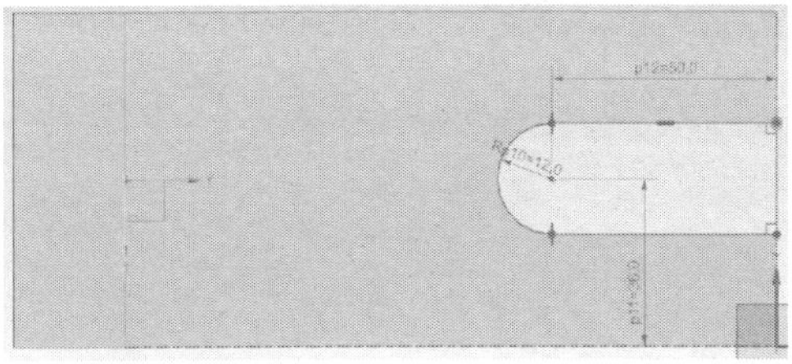

Figure 5-19 Sketch created for the third feature

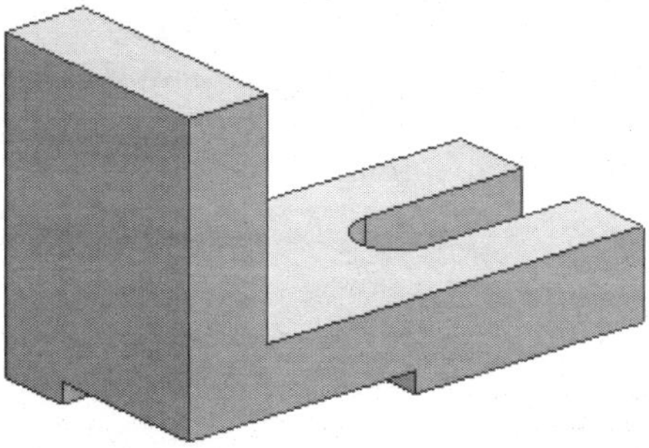

Figure 5-20 Model with the cut feature

Creating the Fourth Feature

The sketch of this feature will be drawn on the front face of the base feature and will be extruded upto a depth of 12 mm by using the **Subtract** option.

1. Invoke the sketching environment by selecting the left face of the base feature as the sketching plane. Make sure that the X-axis of the new sketching plane points toward the right.

2. Draw the sketch of the feature and add the required constraints and dimensions to it, as shown in Figure 5-21.

3. Exit the sketching environment and then invoke the **Extrude** dialog box. Select the sketch.

4. Select the **Subtract** option from the **Boolean** drop-down list if not already selected and reverse the direction of extrusion.

5. Enter **12** in the **Distance** edit box that is available below the **End** drop-down list, and then choose the **OK** button from the dialog box.

6. Hide the sketch. The model with the cut feature is shown in Figure 5-22.

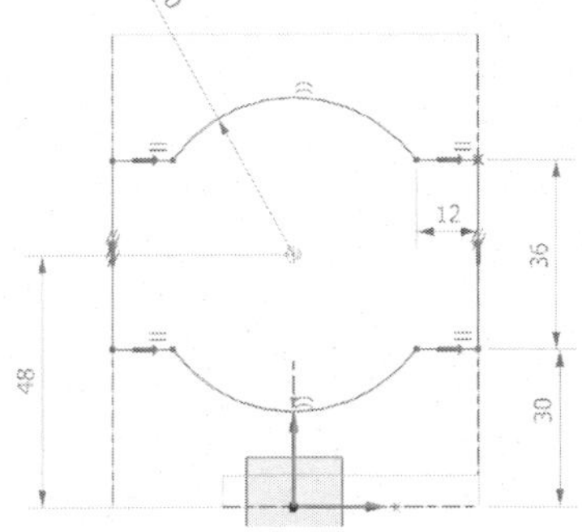

Figure 5-21 Sketch created for the fourth feature

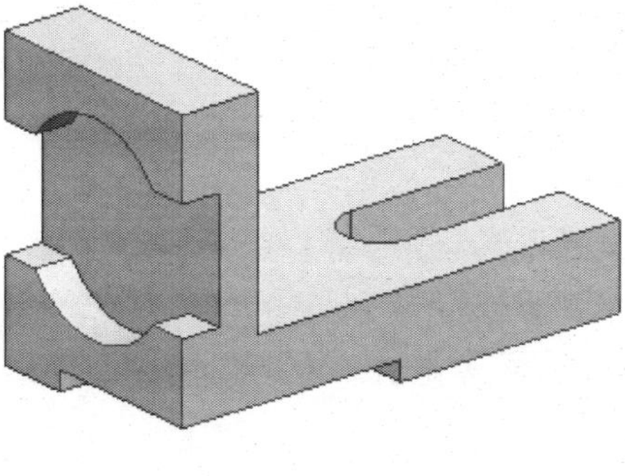

Figure 5-22 Model with the cut feature

Creating the Fifth Feature

1. Similarly, draw a circle of diameter 24 units on the new face that is exposed because of the last cut feature and then extrude the circle by using the **Subtract** option and the **Through All** termination option. Hide the sketch. The final model for Tutorial 1 is shown in Figure 5-23.

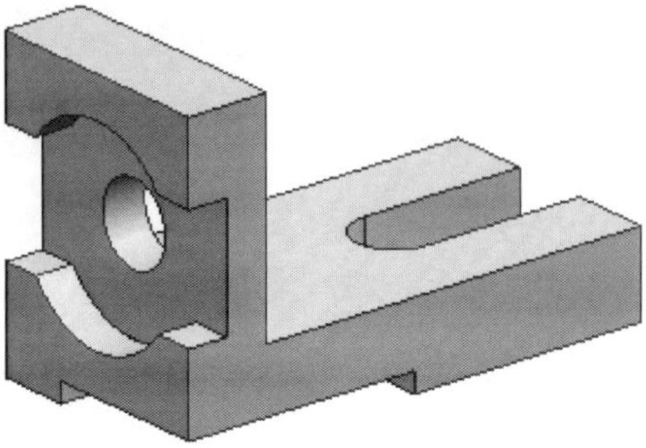

Figure 5-23 *Final model for Tutorial 1*

Saving and Closing the File

1. Choose **Menu > File > Close > Save and Close** from the **Top Border Bar** to save and close the file.

Tutorial 2

In this tutorial, you will create the model shown in Figure 5-24. The dimensions of the model are given in Figure 5-25. **(Expected time: 30 min)**

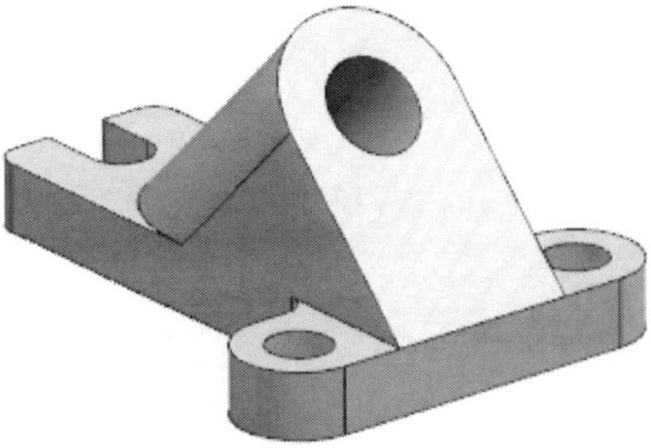

Figure 5-24 *Model for Tutorial 2*

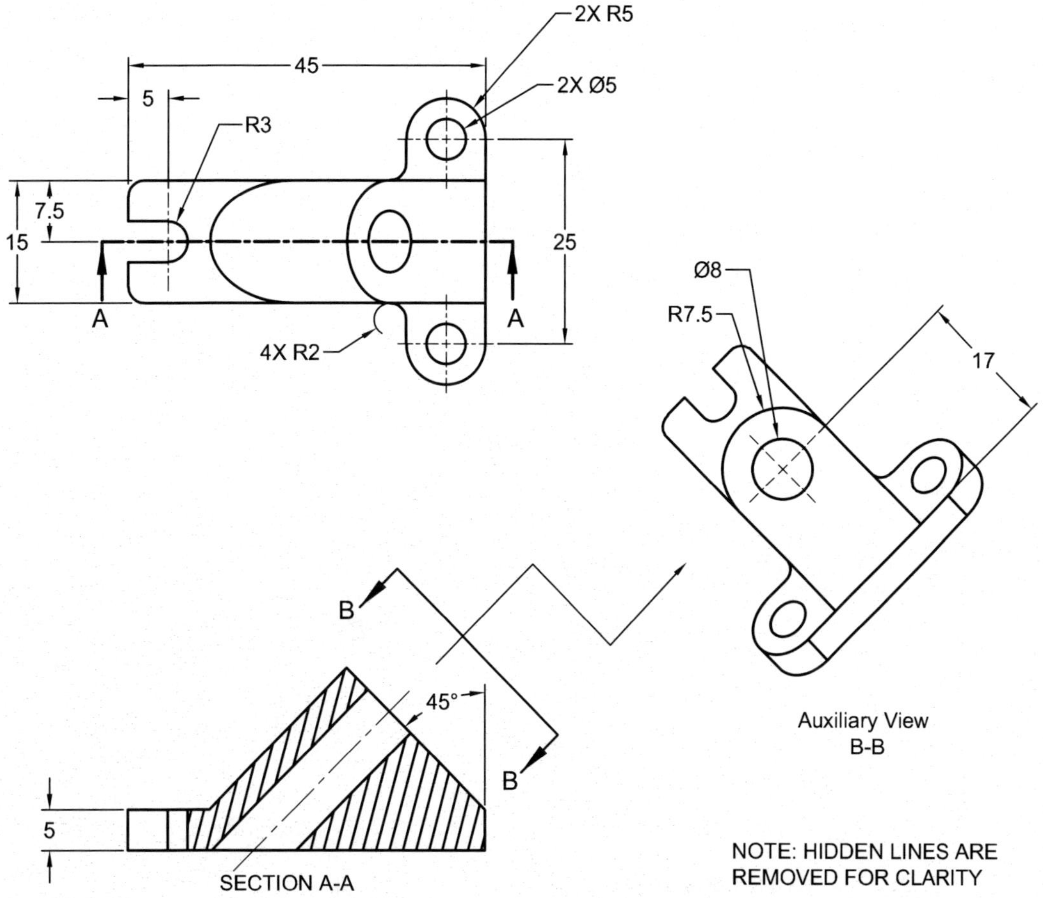

Figure 5-25 *Dimensions of the model for Tutorial 2*

The following steps are required to complete this tutorial:

a. Create the base feature of the model on the XC-YC plane.
b. Create the second feature. In this, you will create a new datum plane at an angle of 45 degrees and draw the sketch of the next extrude feature by using the new datum plane.
c. Create a third feature on the inclined face of the second feature.

Creating the Base Feature

First, you need to create the base feature of the model.

1. Start a new file with the name *c05tut2.prt* using the **Model** template and specify its location as *C:\NX\c05*.

2. Invoke the sketching environment by using the XC-YC plane as the sketching plane and draw a sketch similar to the one shown in Figure 5-26.

3. Exit the sketching environment by choosing the **Finish Sketch** tool.

4. Invoke the **Extrude** dialog box and select the sketch; preview of the extrude feature is displayed.

5. Enter **0** in the **Distance** edit box that is available below the **Start** drop-down list in the **Limits** rollout of the **Extrude** dialog box. Similarly, enter **5** in the **Distance** edit box which is available below the **End** drop-down list. Next, press ENTER.

6. Choose the **OK** button from this dialog box; the base feature of the model is created, refer to Figure 5-27.

7. Hide the sketch and the default datum planes. The base feature after hiding the sketch and the default planes is shown in Figure 5-27.

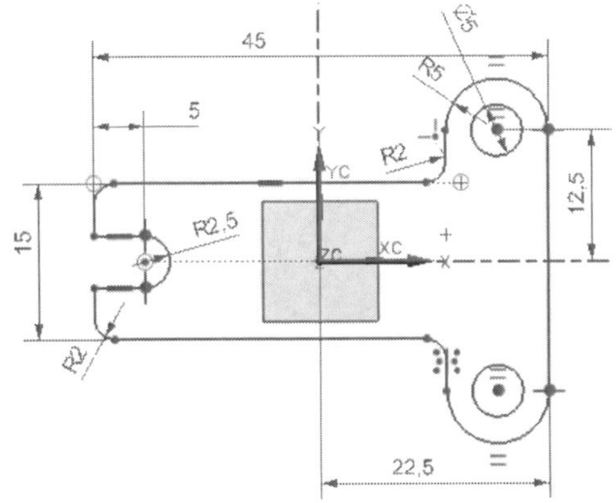

***Figure 5-26** Sketch created for the base feature*

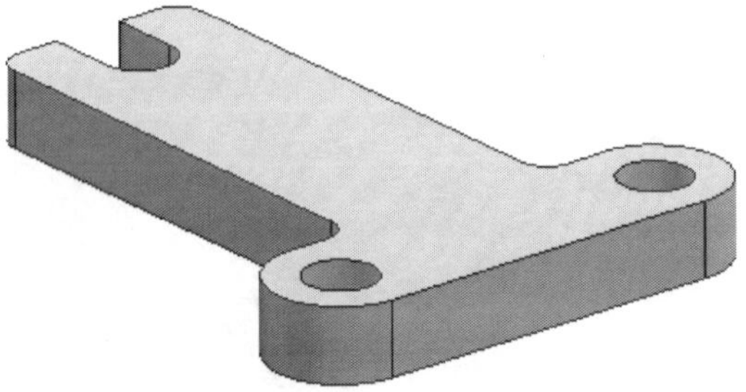

***Figure 5-27** Base feature of the model*

Creating the Second Feature

The second feature is also an extruded feature. However, to create the sketch of this feature, first you need to create a new datum plane at an angle of 45 degrees from the right face of the base feature.

1. Choose the **Datum Plane** tool from the **Feature** group of the **Home** tab; the **Datum Plane** dialog box is displayed.

2. Select the **At Angle** option from the Type drop-down list of this dialog box.

3. First, select the right face of the base feature and then the edge to define the new datum plane, refer to Figure 5-28. On doing so, the **Angle** edit box is displayed.

4. Enter **45** in the **Angle** edit box, as shown in Figure 5-28, and then choose the **OK** button from the **Datum Plane** dialog box.

5. Draw the sketch of the second feature on the newly created datum plane, as shown in Figure 5-29. You need to apply collinear constraints between the lines in the sketch and the edges of the base feature to place the sketch at the desired location.

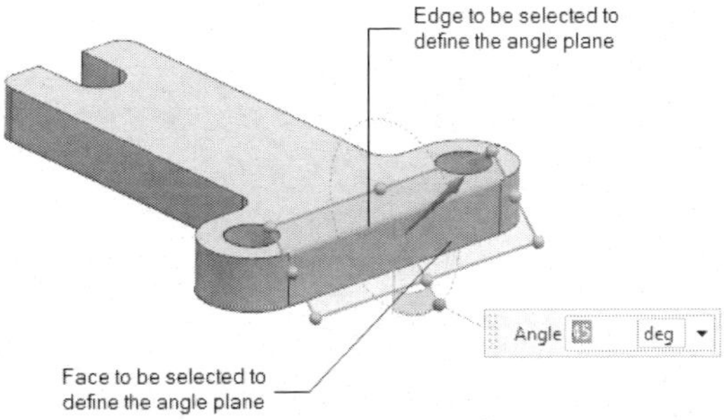

Figure 5-28 *Defining a new datum plane*

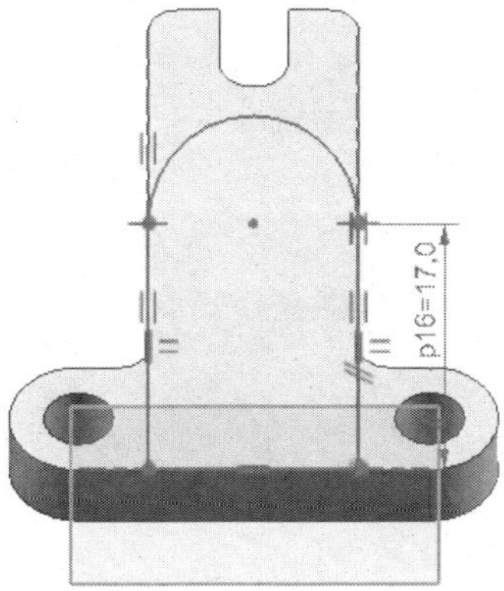

Figure 5-29 *Sketch created for the second feature*

6. Exit the sketching environment and then invoke the **Extrude** dialog box.

7. Select the sketch and then select the **Unite** option from the **Boolean** drop-down list in the dialog box.

8. Select the **Until Next** option from the **End** drop-down list in the **Limits** rollout of the dialog box.

9. Choose the **OK** button from the dialog box; the feature is created.

10. Hide the sketch and the datum plane. Figure 5-30 shows the model after creating the second feature.

Creating the Third Feature

The third feature is an extruded feature and will be created by using the **Subtract** boolean option. The sketch of this feature will be created on the inclined face of the second feature.

1. Select the inclined face of the second feature as the sketching plane and then draw the sketch for the hole. Add the required dimensions and constraints to the circle, refer to Figure 5-25.

2. Exit the sketching environment and then invoke the **Extrude** dialog box.

3. Select the circle and then select the **Subtract** option from the **Boolean** drop-down list in the **Extrude** dialog box.

4. Choose the **Body** button from the **Boolean** rollout if it is not selected by default and then select the existing body from the drawing window; preview of the feature is displayed in the drawing window.

5. Select **Through All** from the **Start** drop-down list in the **Limits** rollout.

6. Choose the **OK** button; the feature is created. Hide the sketch of the circle and then choose the **Fit** tool to fit the model into the screen. The final model for this tutorial is shown in Figure 5-31.

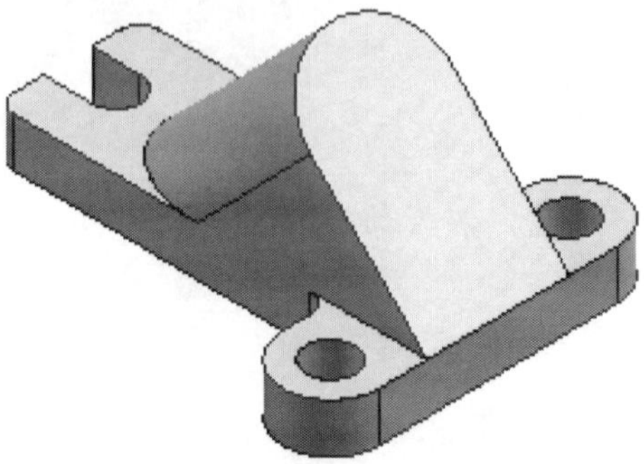

Figure 5-30 *Model after creating the second feature on the inclined datum plane*

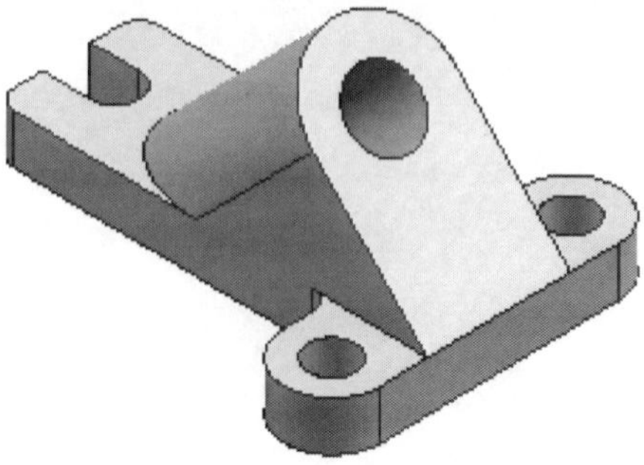

Figure 5-31 *Final model for Tutorial 3*

Saving and Closing the File

1. Choose **Menu > File > Close > Save and Close** from the **Top Border Bar** to save and close the file.

EXERCISES

Exercise 1

Create the model shown in Figure 5-32. The dimensions of the model are shown in Figure 5-32(a). **(Expected time: 30 min)**

Figure 5-32 *Model for Exercise 1*

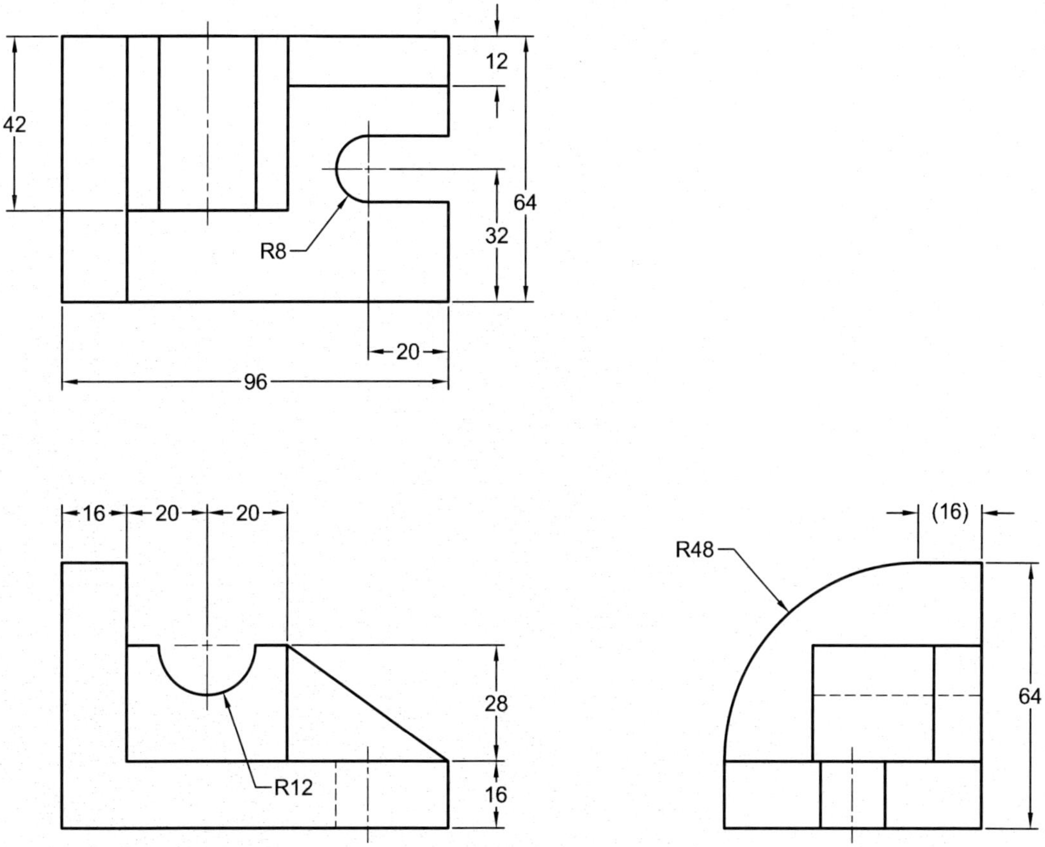

Figure 5-32(a) *Views and dimensions of the model*

Exercise 2

Create the model shown in Figure 5-33. The dimensions of the model are given in Figure 5-33(a). (**Expected time: 30 min**)

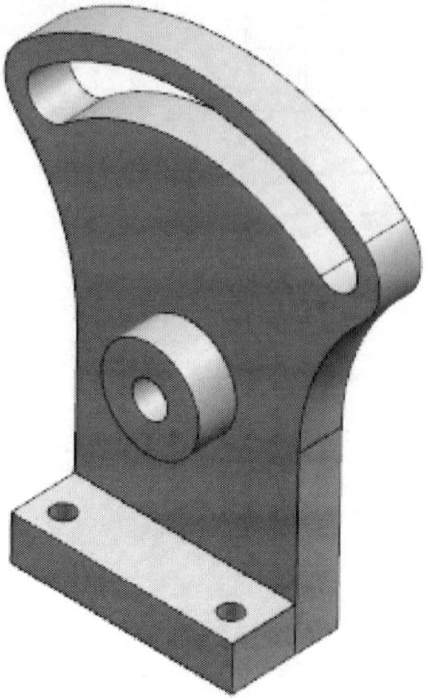

Figure 5-33 *Model for Exercise 2*

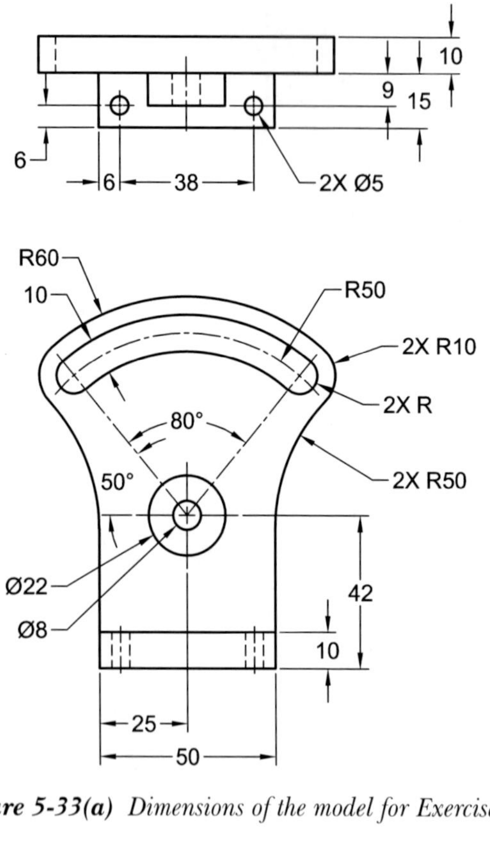

Figure 5-33(a) *Dimensions of the model for Exercise 2*

Chapter 6

Advanced Modeling Tools-I

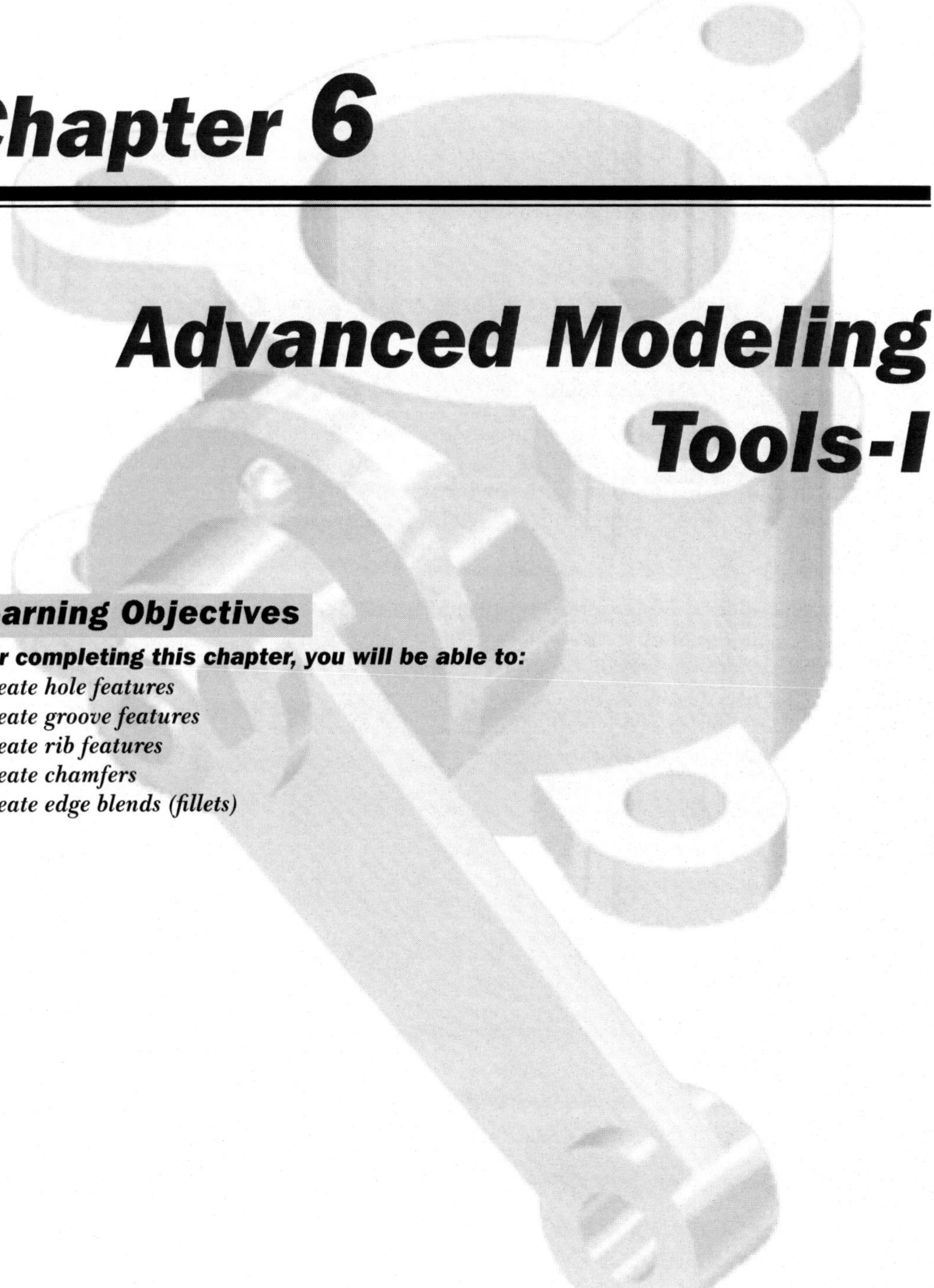

Learning Objectives

After completing this chapter, you will be able to:

- *Create hole features*
- *Create groove features*
- *Create rib features*
- *Create chamfers*
- *Create edge blends (fillets)*

ADVANCED MODELING TOOLS

The advanced modeling tools are mostly used to place different types of standard and user-defined features on a model. Each advanced modeling tool has its specific use in designing a real-world model. Most of the mechanical designs are created using the advanced modeling features such as simple/counterbore/countersink holes, pocket, groove, slot, dart, and so on. NX provides you with a number of advanced modeling tools that assist in creating these advanced features. Note that these features are parametric in nature and can be modified or edited any time. The advanced feature tools are used to reduce the time taken in creating the design. Also, you can create and save user-defined features in NX. These features can be placed any number of times on a model.

CREATING HOLES BY USING THE HOLE TOOL

Ribbon:	Home > Feature > Hole
Menu:	Insert > Design Feature > Hole

The **Hole** tool is used to create a through, blind cylindrical, or conical cutout in a model. A hole can be threaded or non-threaded. To invoke this tool, choose **Home > Feature > Hole** from the **Ribbon**; the **Hole** dialog box will be displayed, as shown in Figure 6-1.

By default, the **General Hole** option is selected in the Type drop-down list and the **Simple** option is selected in the **Form** drop-down list. As a result, you are prompted to select a planar face to sketch or specify points. You can create different types of holes such as counterbore, drilled, countersunk, threaded and so on, depending on the option selected.

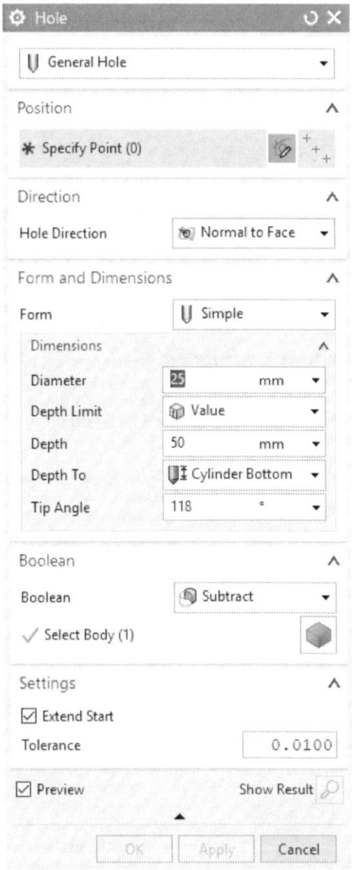

*Figure 6-1 The **Hole** dialog box*

CREATING GROOVES

Ribbon: Home > Feature > Design Feature Drop-down > Groove *(Customize to Add)*
Menu: Insert > Design Feature > Groove

Grooves are the channels that are created on the outer surface of a cylindrical part or a conical feature. The groove operation can only be performed on the features that have their own center axis. To create grooves, choose the **Groove** tool from the **Design Feature Drop-down** of the **Feature** group in the **Home** tab; the **Groove** dialog box will be displayed, as shown in Figure 6-2. Also, you will be prompted to choose the required groove type. In NX, you can create three types of grooves, refer to Figure 6-2.

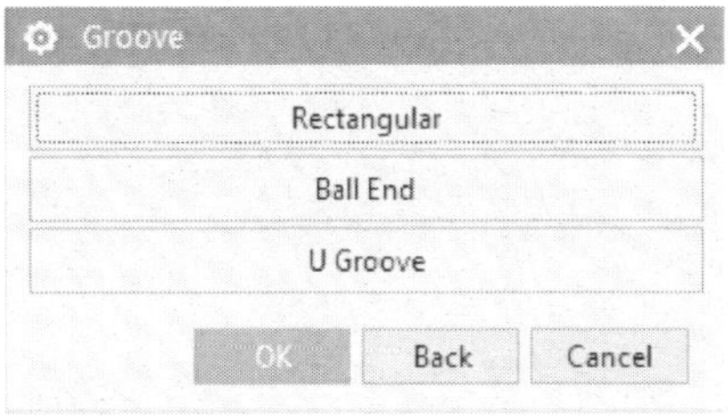

***Figure 6-2** The **Groove** dialog box*

CREATING RIBS

Ribbon: Home > Feature > Design Feature Drop-down > Rib *(Customize to add)*
Menu: Insert > Design Feature > Rib

Ribs are defined as the thin-walled structures that are used to increase the strength of the entire structure of a component so that it does not fail under an increased load. In NX, the ribs are created using an open curve as well as a closed curve. To create a rib feature, choose the **Rib** tool from the **Design Feature Drop-down** of the **Feature** group in the **Home** tab; the **Rib** dialog box will be displayed, refer to Figure 6-3. In this dialog box, by default, the solid body will be selected as the target body. Also, you will be prompted to select a curve or create a curve to create the rib feature. Select the curve from the graphic area. Next, you need to enter the thickness value of thin wall in the **Thickness** edit box of the **Walls** rollout. After specifying values, choose the **OK** button to create the rib.

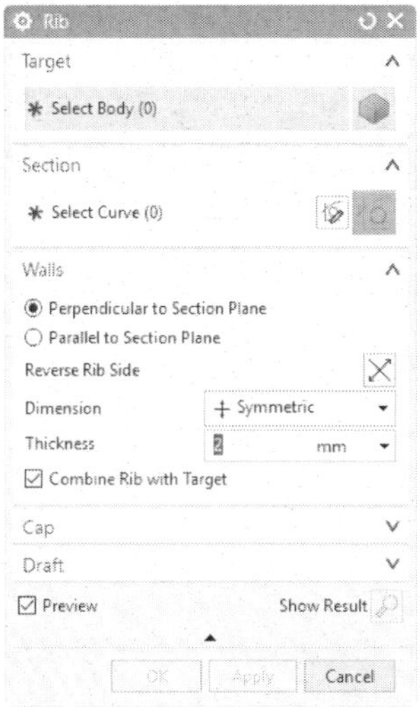

***Figure 6-3** The **Rib** dialog box*

CREATING CHAMFERS

Ribbon: Home > Feature > Chamfer
Menu: Insert > Detail Feature > Chamfer

A chamfer feature is created on the sharp corners of a model to reduce stress concentration and to prevent accidental injuries due to sharp corners. To create a chamfer feature, choose the **Chamfer** tool from the **Feature** group of the **Home** tab; the **Chamfer** dialog box will be displayed, refer to Figure 6-4. In NX, there are three methods to create a chamfer feature: Symmetric, Asymmetric, and Offset and Angle.

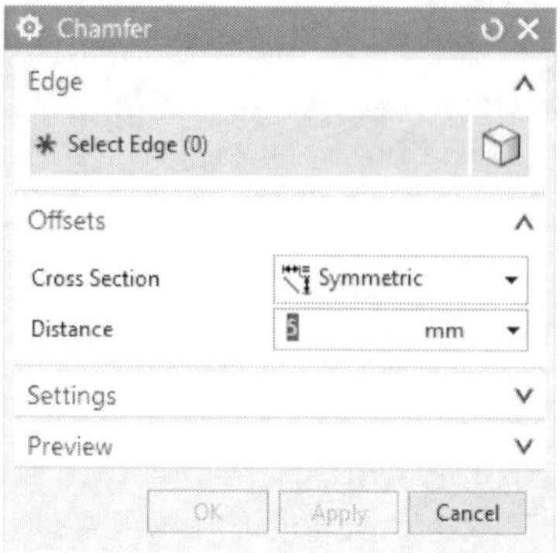

*Figure 6-4 The **Chamfer** dialog box*

CREATING AN EDGE BLEND

Ribbon:	Home > Feature > Blend Drop-down > Edge Blend
Menu:	Insert > Detail Feature > Edge Blend

An edge blend is a fillet operation performed on the sharp corners of a model. In NX, you can create different types of fillets using this tool. To create an edge blend, choose the **Edge Blend** tool from the **Blend Drop-down** of the **Feature** group of the **Home** tab; the **Edge Blend** dialog box will be displayed, as shown in Figure 6-5. By default, the **Circular** option is selected in the **Shape** drop-down list of the **Edge** rollout in the dialog box and you are prompted to select the edges to fillet. The options in the various rollouts of the **Edge Blend** dialog box are discussed next.

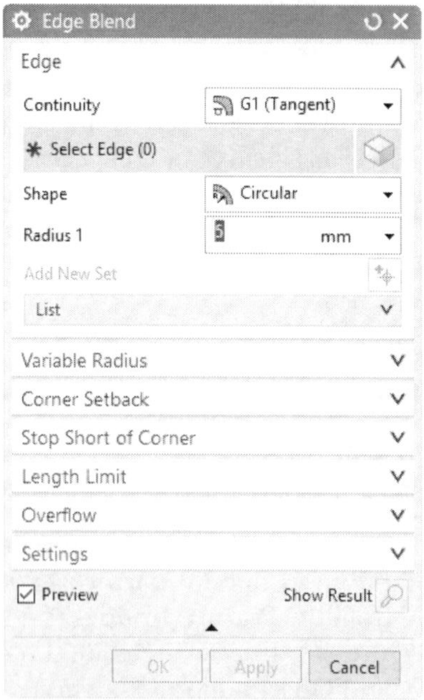

*Figure 6-5 The **Edge Blend** dialog box*

TUTORIALS

Tutorial 1

In this tutorial, you will create the model shown in Figure 6-6(a). The dimensions of the model are shown in Figure 6-6(b). After creating the model, save it with the name *c06tut1.prt* at the following location: *\NX\c06*. **(Expected time: 30 min)**

Figure 6-6(a) *Model for Tutorial 1*

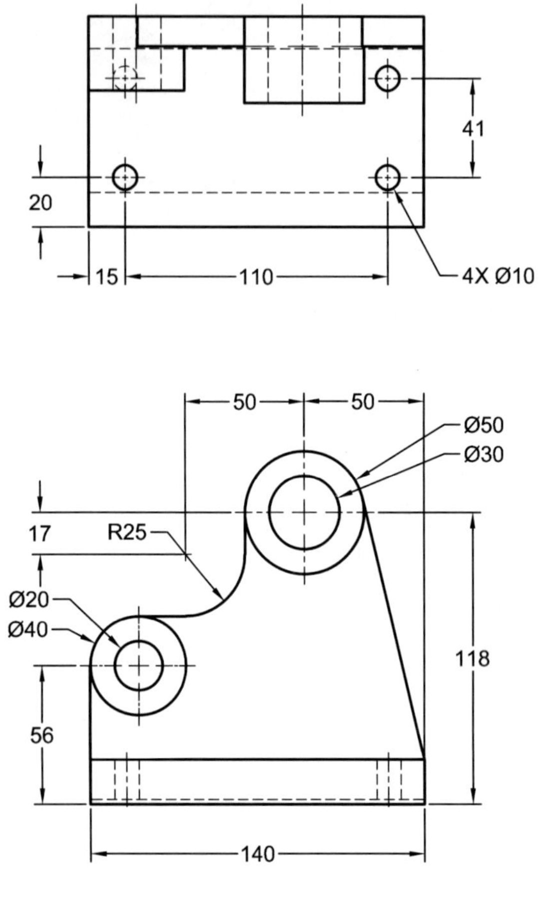

Figure 6-6(b) *Views and dimensions of the model*

The following steps are required to complete this tutorial:

a. Draw the sketch for the base feature and extrude it, refer to Figures 6-7 and 6-8.
b. Draw the sketch for the second feature and extrude it, refer to Figures 6-9 and 6-10.
c. Draw the sketch for the third feature and extrude it, refer to Figures 6-11 and 6-12.
d. Draw the sketch for the fourth feature and extrude it, refer to Figures 6-13 and 6-14.
e. Create two holes on the side face of the model using the **Hole** tool, refer to Figures 6-15 and 6-16.
f. Create four holes on the top face of the base feature using the **Hole** tool, refer to Figures 6-18 through 6-19.
g. Save and close the file.

Creating the Base Feature of the Model

The base feature will be created using the profile of the right face of the model.

1. Start a new file with the name *c06tut1.prt* using the **Model** template and specify its location as *C:\NX\c06*.

2. Draw the sketch by using the YC-ZC plane as shown in Figure 6-7.

3. Choose the **Finish Sketch** tool from the **Direct Sketch** group to exit the Sketch environment.

4. Choose the **Extrude** tool from the **Feature** group; the **Extrude** dialog box is displayed and you are prompted to select the planar face to sketch or select the section geometry. Select the sketch created for the base feature from the drawing window.

5. Select the **Symmetric Value** option from the **Start** drop-down list and enter **70** in the **Distance** edit box.

6. Choose the **Apply** button and then close the dialog box by choosing the **Cancel** button. The resulting base feature is shown in Figure 6-8.

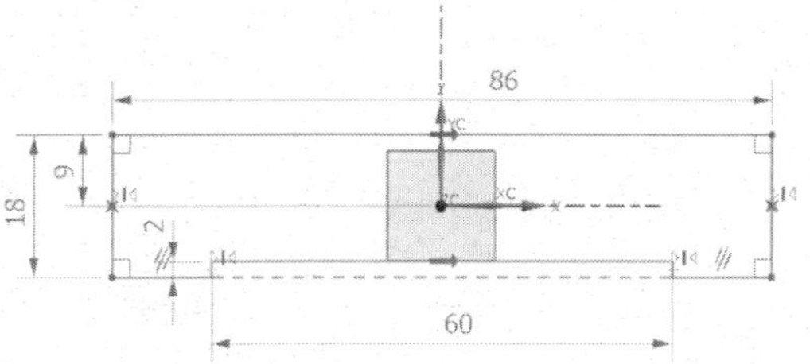

Figure 6-7 *Sketch for the base feature*

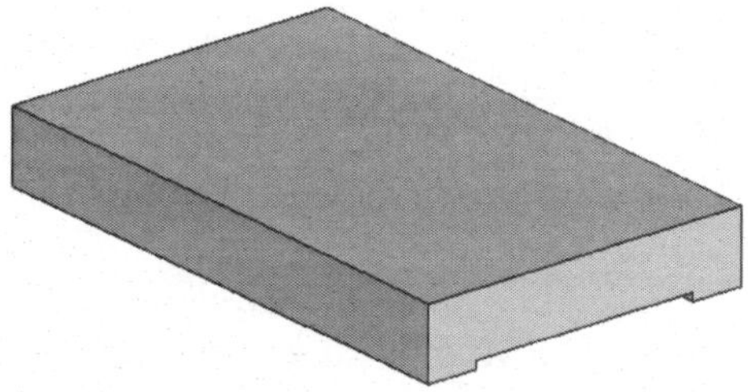

Figure 6-8 *The resulting base feature*

Creating the Second Feature

Next, you need to create the second feature using the steps given next.

1. Invoke the **Create Sketch** dialog box by choosing the **Sketch** tool from the **Direct Sketch** group. Next, select the back face of the base feature as the sketching plane.

2. Draw the sketch for the second feature, refer to Figure 6-9.

3. Choose the **Finish Sketch** tool from the **Direct Sketch** group to exit the sketching environment.

4. Choose the **Extrude** tool from the **Feature** group; you are prompted to select the planar face to sketch or select the section geometry.

5. Select the sketch created for the second feature. Next, in the **Limits** rollout of the dialog box enter **0** and **12** in the **Distance** edit boxes available below the **Start** and **End** drop-down lists, respectively. Make sure the sketch is extruded in the front direction.

6. Select the **Unite** option from the **Boolean** drop-down list in the **Boolean** rollout of the dialog box.

7. Choose the **Apply** button and then close the dialog box by choosing the **Cancel** button. The resulting model after creating the second feature is shown in Figure 6-10.

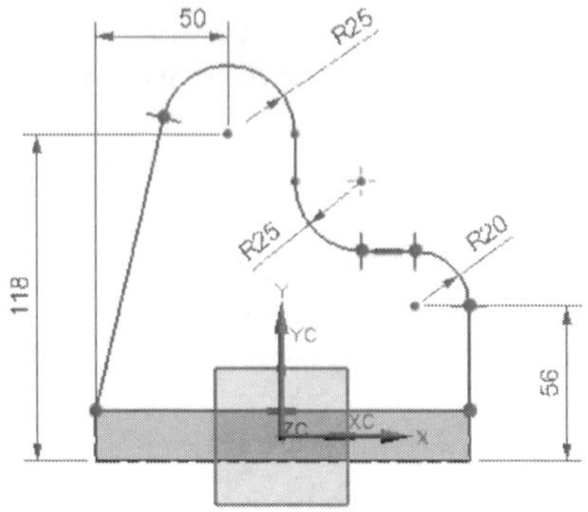

Figure 6-9 Sketch for the second feature

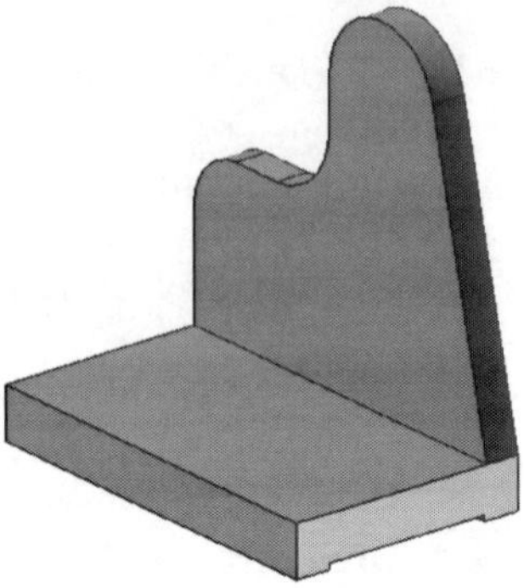

Figure 6-10 Model after creating the second feature

Creating the Third Feature

Now, you need to create the third feature, which is an extrude feature.

1. Invoke the **Create Sketch** dialog box by choosing the **Sketch** tool from the **Direct Sketch** group. Next, select the front face of the second feature as the sketching plane.

2. Draw the sketch for the third feature which is a circle of 50 mm diameter, as shown in Figure 6-11.

3. Choose the **Finish Sketch** tool from the **Direct Sketch** group to exit the sketching environment.

4. Choose the **Extrude** tool from the **Feature** group; the **Extrude** dialog box is displayed and you are prompted to select the planar face to sketch or select the section geometry.

5. Select the sketch created for the third feature and enter **23** in the **Distance** edit box below the **End** drop-down list in the **Limits** rollout of the dialog box. Make sure the sketch is being extruded toward the screen.

6. Select the **Unite** option from the **Boolean** drop-down list in the **Boolean** rollout of the dialog box.

7. Choose the **OK** button from the dialog box. The model after creating the third feature is shown in Figure 6-12.

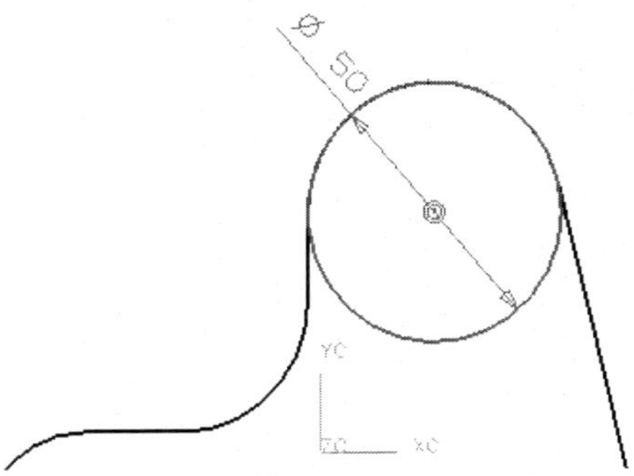

Figure 6-11 *Sketch for the third feature*

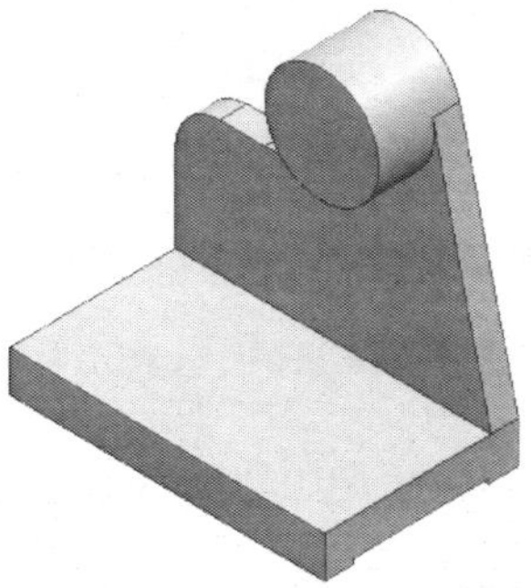

Figure 6-12 *Model after creating the third feature*

Creating the Fourth Feature

The steps given next are required to create the fourth feature.

1. Invoke the sketching environment by selecting the front face of the second feature as the sketching plane.

2. Draw the sketch for the fourth feature which is a circle of 40 mm diameter, as shown in Figure 6-13.

3. Choose the **Finish Sketch** tool from the **Direct Sketch** group to exit the sketching environment.

4. Choose the **Extrude** tool from the **Feature** group; the **Extrude** dialog box is displayed and you are prompted to select the planar face to sketch or select the section geometry.

5. Select the sketch created for the fourth feature and enter **18** in the **Distance** edit box available below the **End** drop-down list. Make sure the sketch is extruded toward the front.

6. Select the **Unite** option from the **Boolean** drop-down list in the **Boolean** rollout of the dialog box.

7. Choose the **OK** button from the dialog box. The resulting model after creating the fourth feature is shown in Figure 6-14.

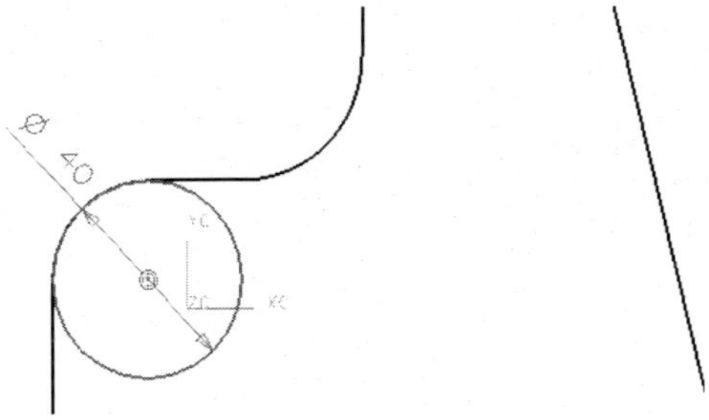

Figure 6-13 *Sketch for the fourth feature*

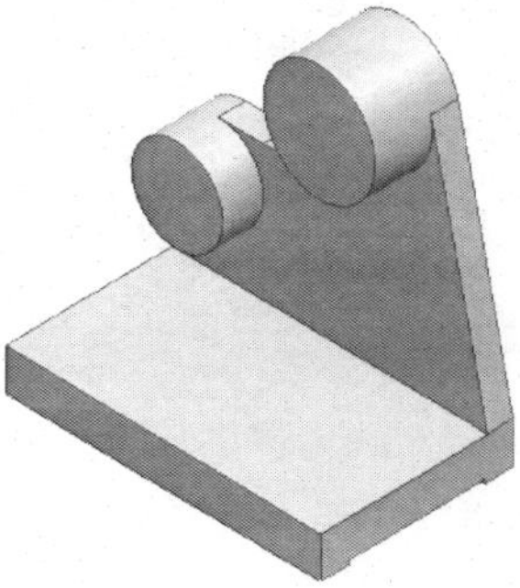

Figure 6-14 *Model after creating the fourth feature*

Creating a Hole in the Third Feature

Follow the steps given next to create a hole in the third feature.

1. Choose the **Hole** tool from the **Feature** group; the **Hole** dialog box is displayed and you are prompted to select the planar face to sketch or specify points.

2. Select the **General Hole** option from the drop-down list in the type drop-down list and the **Simple** option from the **Form** drop-down list in the **Form and Dimensions** rollout of the dialog box if they are not selected by default.

3. Move the cursor over the center point of the circular face of the third feature. Click the left mouse button when the cursor snaps to the center point and the coincident constraint is displayed below the cursor; preview of the hole feature is displayed in the drawing window.

4. Enter **30** in the **Diameter** edit box of the **Dimensions** sub-rollout.

5. Select the **Through Body** option from the **Depth Limit** drop-down list in the **Dimensions** sub-rollout of the **Form and Dimensions** rollout. Next, select the **Subtract** option from the **Boolean** drop-down list.

6. Accept the other default settings and choose the **OK** button from the dialog box; a hole is created in the third feature, as shown in Figure 6-15.

Creating a Hole in the Fourth Feature

In this section, you will create a hole in the fourth feature using the following steps:

1. Choose the **Hole** tool from the **Feature** group; the **Hole** dialog box is displayed and you are prompted to select a planar face to sketch or specify points.

2. Move the cursor over the center point of the circular face of the fourth feature; the center point is highlighted. Now, select it; the preview of the hole is displayed in the drawing window.

3. Enter **20** in the **Diameter** edit box in the **Dimensions** sub-rollout.

4. Select the **Through Body** option from the **Depth Limit** drop-down list in the **Dimensions** sub-rollout of the dialog box.

5. Accept other default settings and choose the **OK** button from the **Hole** dialog box; a hole is created in the fourth feature, as shown in Figure 6-16.

Figure 6-15 Model after creating a hole in the third feature

Figure 6-16 Model after creating a hole in the fourth feature

Creating Four Holes in the Base Feature

To complete the model, you need to create four holes in the base feature by using the steps discussed next.

1. Choose the **Hole** tool from the **Feature** group; the **Hole** dialog box is displayed and you are prompted to select a planar face to sketch or specify points.

2. Select the top face of the base feature as the planar placement plane; the **Sketch Point** dialog box is displayed and you are prompted to define a point. Also, the Sketch in Task environment is invoked and a point is placed on the top face of base feature.

3. Choose the **Close** button from the **Sketch Point** dialog box. Now, apply the dimension to the point, as shown in Figure 6-17.

4. Choose the **Finish** tool from the **Sketch** group of the **Home** tab to exit the Sketch in Task environment. As you exit the Sketch in Task environment, the **Hole** dialog box is displayed again. Also, the preview of the hole is displayed. Note that the center of the hole is at the point that you have specified using the **Sketch Point** dialog box.

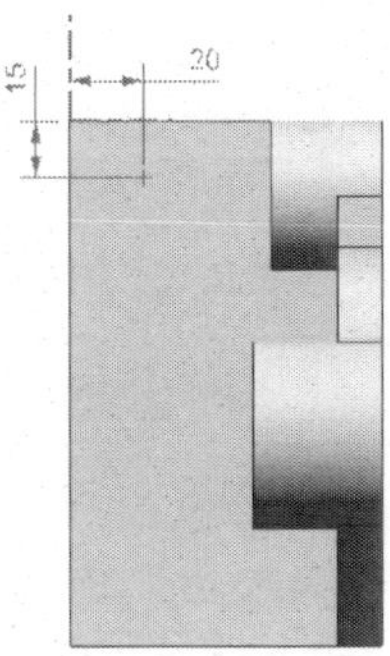

Figure 6-17 Point after applying the dimension

5. Enter **10** as the diameter of the hole in the **Diameter** edit box of the **Dimensions** sub-rollout of the **Form and Dimensions** rollout in the **Hole** dialog box.

6. Select the **Through Body** option from the **Depth Limit** drop-down list in the **Dimensions** sub-rollout of the **Form and Dimensions** rollout of the dialog box.

7. Accept other default settings and choose the **OK** button from the dialog box; a hole is created in the base feature, as shown in Figure 6-18.

8. Similarly, create the remaining three holes. Refer to Figure 6-6(b) for dimensions. The final model for this tutorial is shown in Figure 6-19.

Figure 6-18 *Model after creating the first hole in the base feature*

Figure 6-19 *Final model after creating all the four holes in the base feature*

Saving and Closing the File

1. Choose **Menu > File > Close > Save and Close** from the **Top Border Bar** to save and close the file.

Tutorial 2

In this tutorial, you will create the model shown in Figure 6-20. The drawing views and the dimensions are shown in Figure 6-21. After creating the model, save it with the name *c06tut2.prt* at the following location: *\NX\c06.* **(Expected time: 30 min)**

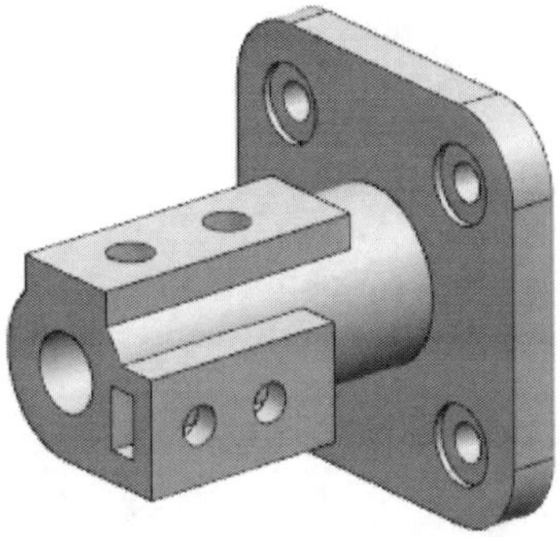

Figure 6-20 *Model for Tutorial 2*

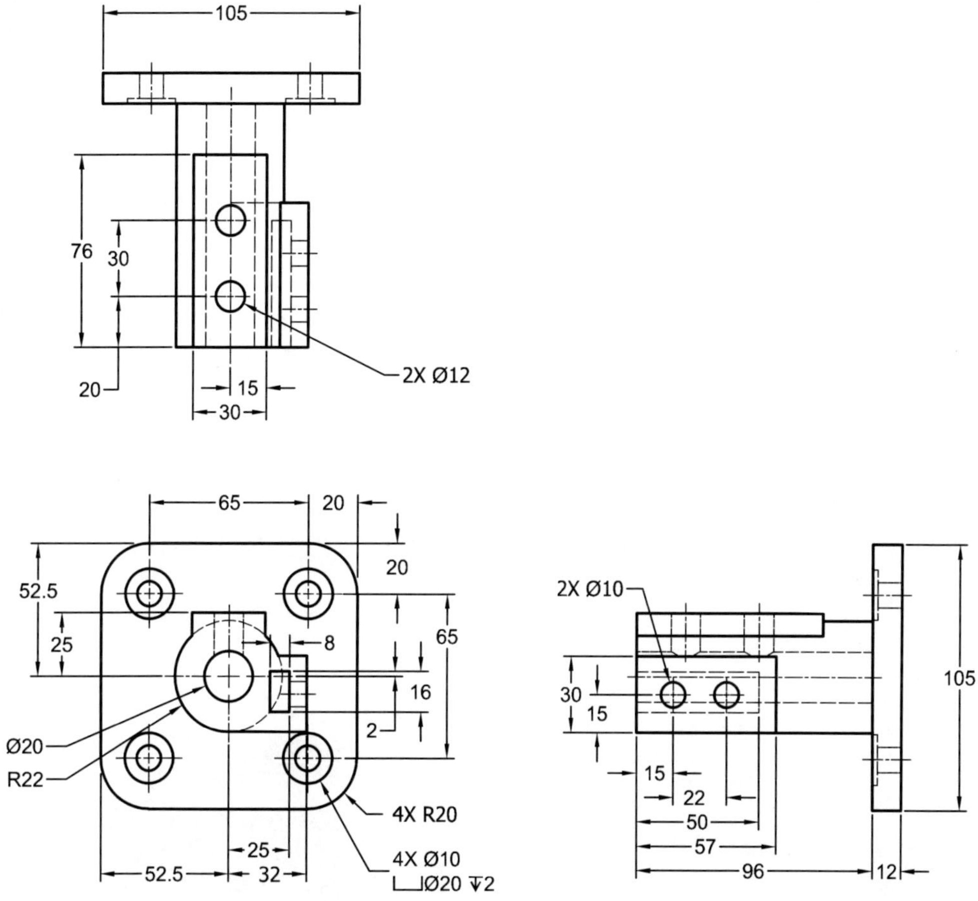

***Figure 6-21** The drawing views and the dimensions of the model for Tutorial 2*

The following steps are required to complete this tutorial:

a. Draw the sketch of the base feature of the model and extrude it, refer to Figures 6-22 and 6-23.
b. Draw the sketch of the second feature, which is a circle, and extrude it, refer to Figures 6-24 and 6-25.
c. Draw the sketches of the third and fourth features and extrude them, refer to Figures 6-26 and 6-27.
d. Create the simple hole on the left face of the model using the **Hole** tool, refer to Figure 6-28.
e. Create the rectangular cut extrude feature, refer to Figure 6-29.
f. Create holes on the top and front faces of the model using the **Hole** tool, refer to Figures 6-30 to 6-33.
g. Create four counterbore holes in the base feature using the **Hole** tool, refer to Figures 6-34 and 6-35.
h. Save and close the file.

Creating the Base Feature of the Model

After starting a new file, you need to create the sketch of the base feature of the model by following the steps given next.

1. Start a new file with the name *c06tut2.prt* using the **Model** template and specify its location as *C:\NX\c06*.

2. Draw the sketch of the base feature by using the XC-ZC plane as the sketching plane, as shown in Figure 6-22. Next, exit the sketching environment.

3. Choose the **Extrude** tool from the **Feature** group; the **Extrude** dialog box is displayed and you are prompted to select the sketch to be extruded. Select the sketch created for the base feature.

4. Enter **12** in the **Distance** edit box available below the **End** drop-down list in the **Limits** rollout of the **Extrude** dialog box.

5. Choose the **OK** button from the **Extrude** dialog box; the base feature is created, as shown in Figure 6-23.

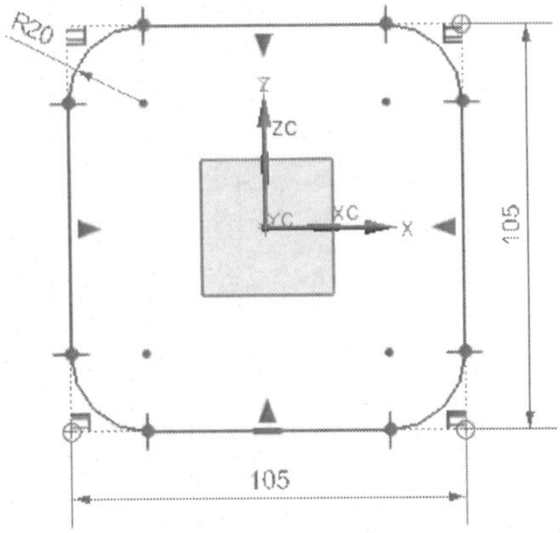

Figure 6-22 *The sketch of the base feature*

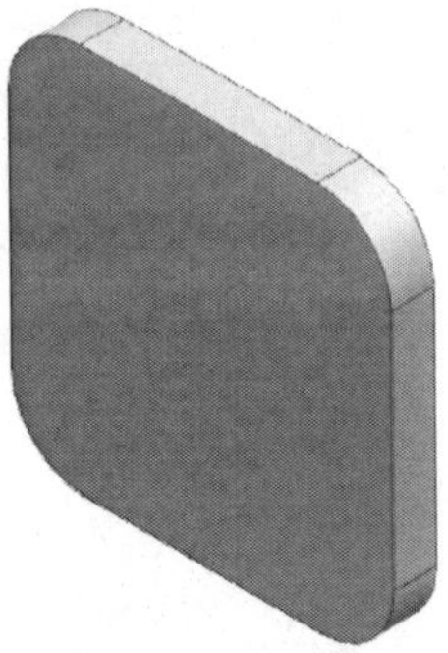

Figure 6-23 *The base feature*

Creating the Second Feature

The second feature of the model is a cylindrical extrude feature. The following steps explain the procedure to create this feature:

1. Invoke the sketching environment by selecting the front face of the base feature as the sketching plane.

2. Draw the sketch for the second feature which is a circle of diameter 44, refer to Figure 6-24. Next, choose the **Finish Sketch** tool from the **Direct Sketch** group to exit the sketching environment.

3. Choose the **Extrude** tool from the **Feature** group; the **Extrude** dialog box is displayed and you are prompted to select a sketch.

4. Select the sketch created for the second feature and enter **96** in the **Distance** edit box available below the **End** drop-down list in the **Limits** rollout of the **Extrude** dialog box.

5. Select the **Unite** option from the **Boolean** drop-down list in the **Boolean** rollout.

6. Choose the **OK** button from the dialog box; the cylindrical extrude feature is created. The model after creating the second feature is shown in Figure 6-25.

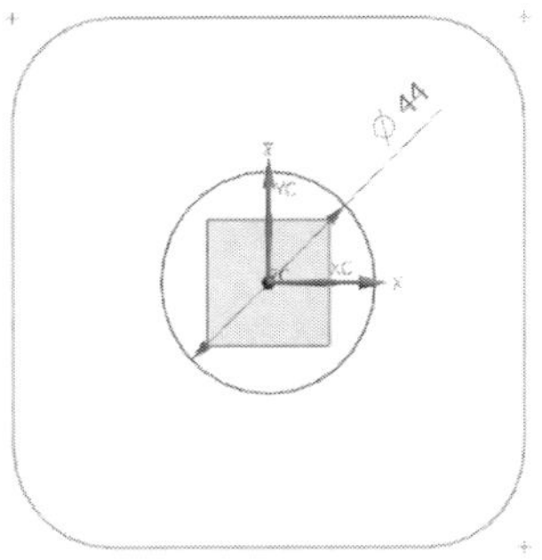

Figure 6-24 *The sketch for the second feature*

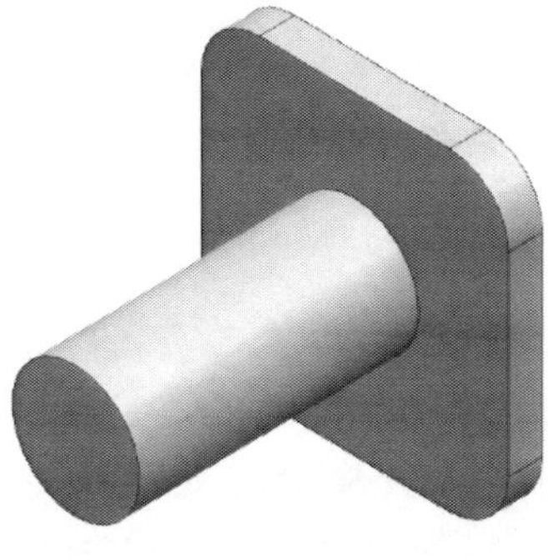

Figure 6-25 *Model after creating the second feature*

Creating the Third and Fourth Features

In this section, you need to create the third and fourth features which are extrude features, by following the steps given next.

1. Invoke the sketching environment by selecting the front face of the second feature as the sketching plane.

2. Draw the rectangular sketch for the third feature by using the dimensions given in Figure 6-21.

3. Choose the **Finish Sketch** tool from the **Direct Sketch** group to exit the sketching environment.

4. Choose the **Extrude** tool from the **Feature** group; the **Extrude** dialog box is displayed. Select the sketch created for the third feature; a preview of the feature is displayed in the drawing window.

5. Reverse the direction of extrusion by choosing the **Reverse Direction** button from the **Direction** rollout of the dialog box.

6. Enter **76** as the depth value in the **Distance** edit box available below the **End** drop-down list of the **Limits** rollout in the dialog box. Also, select the **Unite** option from the **Boolean** drop-down list in the **Boolean** rollout. Make sure that the value in the **Distance** edit box available below the **Start** drop-down list is 0.

7. Choose the **OK** button from the dialog box; the extruded feature is created. The resulting model after creating the third feature is shown in Figure 6-26.

8. Similarly, create the fourth extruded feature by using the dimensions given in Figure 6-21. The model after creating the fourth feature is shown in Figure 6-27.

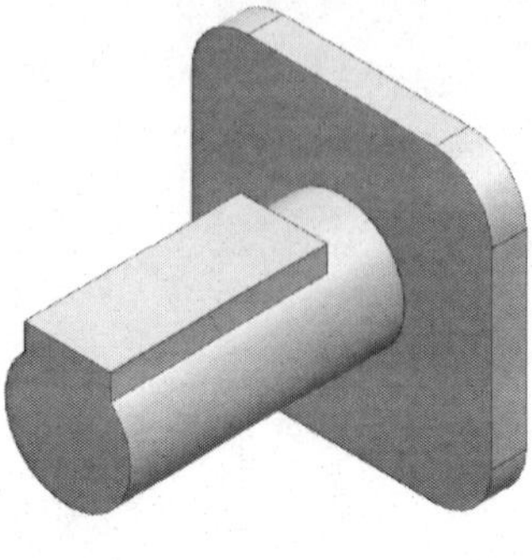

Figure 6-26 *The resulting model after creating the third feature*

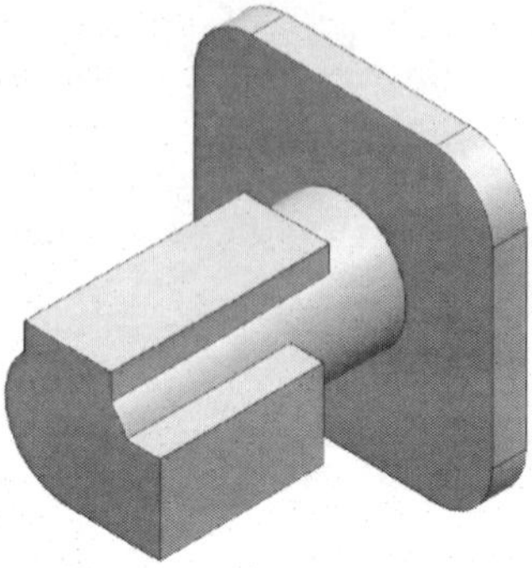

Figure 6-27 *The model after creating the fourth feature*

Creating a Hole in the Second Feature

In this section, you need to create a hole in the second feature.

1. Choose the **Hole** tool from the **Feature** group; the **Hole** dialog box is displayed and you are prompted to specify a point.

2. Select the center point of the front circular face of the second feature as the placement point; a preview of the hole feature is displayed in the drawing window. Make sure that the **General Hole** and **Simple** options are selected in the type and **Form** drop-down lists, respectively in their rollouts.

3. Enter **20** in the **Diameter** edit box. Next, select the **Value** option from the **Depth Limit** drop-down list and enter **96** in the **Depth** edit box. Also, enter **0** in the **Tip Angle** edit box in the **Dimensions** sub-rollout of the **Hole** dialog box.

4. Accept other default settings and choose the **OK** button from the dialog box; a hole is created, as shown in Figure 6-28.

Creating the Rectangular Cut Extrude Feature

In this section, you need to create the rectangular cut extrude feature by following the steps given next.

1. Invoke the sketching environment by selecting the front face of the fourth feature which is merged with the front face of the second feature as the sketching plane.

2. Draw the sketch of this feature by using the dimensions given in Figure 6-21. Choose the **Finish Sketch** tool from the **Direct Sketch** group to exit the sketching environment.

3. Choose the **Extrude** tool from the **Feature** group and select the sketch created for the feature; a preview of the feature is displayed in the drawing window. Reverse the direction of feature creation by using the **Reverse Direction** button, if required.

4. Enter the depth value **50** in the **Distance** edit box available below the **End** drop-down list. Make sure the value in the **Distance** edit box available below the **Start** drop-down list is 0. Next, select the **Subtract** option from the **Boolean** drop-down list in the **Boolean** rollout.

5. Choose the **OK** button from the dialog box; the cut extrude feature is created. The resulting model after creating this feature is shown in Figure 6-29.

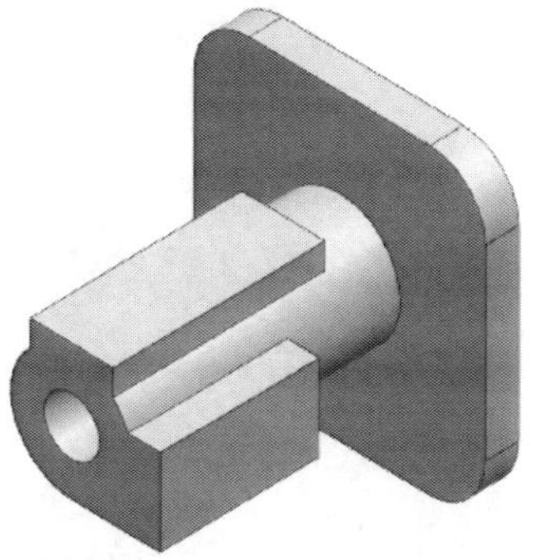

Figure 6-28 *Model after creating the hole feature*

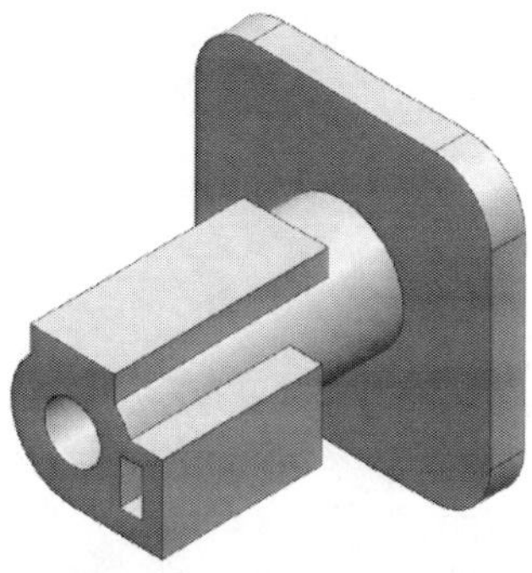

Figure 6-29 *Model after creating the rectangular cut feature*

Creating Holes in the Third and Fourth Features

In this section, you need to create holes in the third and fourth features by following the steps given next.

1. Choose the **Hole** tool from the **Feature** group; the **Hole** dialog box is displayed and you are prompted to select a planar face to sketch or specify a point. Make sure the **General Hole** and **Simple** options are selected in the type and **Form** drop-down lists, respectively.

2. Select the top planar face of the third feature as a placement face; the **Sketch Point** dialog box is displayed. Also, the Sketch in Task environment is invoked. Now, place the point anywhere on the face and then exit the **Sketch Point** dialog box.

3. Apply dimensions to the point, refer to Figure 6-30. Next, exit the Sketch in Task environment; the **Hole** dialog box is displayed. Also, a preview of the hole is displayed in the drawing window.

4. Enter **12** as the hole diameter in the **Diameter** edit box in the **Dimensions** sub-rollout of the dialog box.

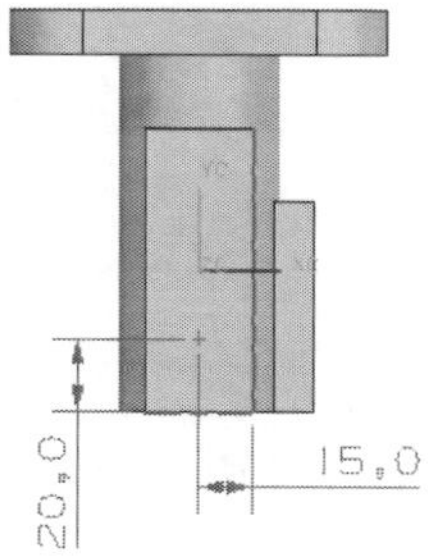

Figure 6-30 *Point after applying the dimensions*

5. Select the **Until Selected** option from the **Depth Limit** drop-down list in the **Dimensions** sub-rollout of the dialog box; you are prompted to select a face or a datum plane on which the hole is to be created.

6. Select the cylindrical face of the hole created on the second feature as the face on which the hole is to be created.

7. Accept other default settings and choose the **OK** button from the dialog box; a hole is created, as shown in Figure 6-31.

8. Similarly, create the second hole on the third feature of the model using the **Hole** dialog box. The model after creating the second hole is shown in Figure 6-32.

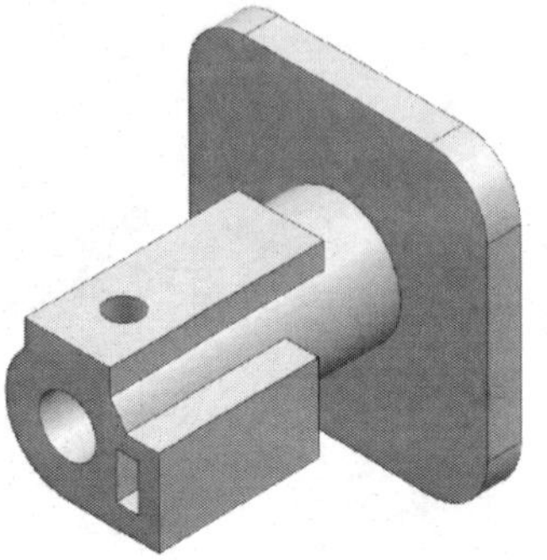

Figure 6-31 *Model after creating the first hole*

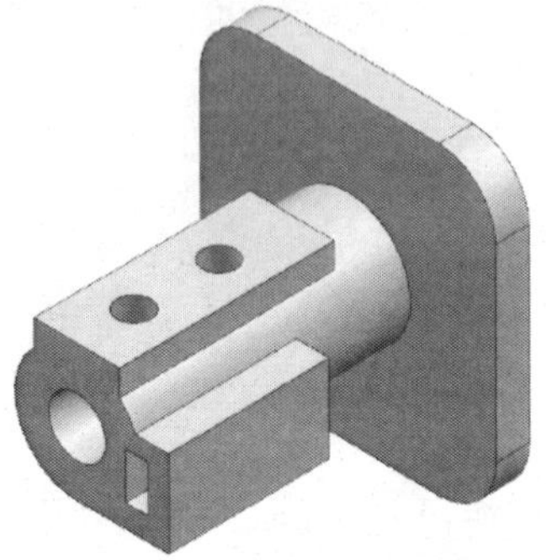

Figure 6-32 *Model after creating both holes*

After creating holes on the third feature, create holes on the fourth feature of the model by using the **Hole** dialog box. These holes will be created on the right face of the fourth feature. For dimensions of the holes, refer to Figure 6-21. The model after creating the holes in the fourth feature is shown in Figure 6-33.

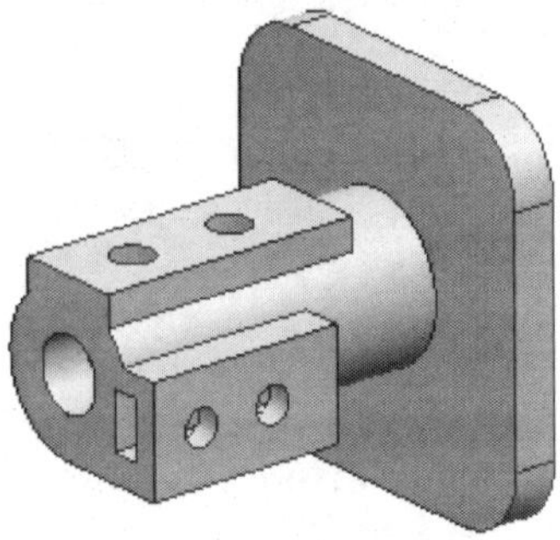

Figure 6-33 *Model after creating holes in the fourth feature*

Creating Counterbore Holes in the Base Feature

Next, create four counterbore holes in the base feature by following the steps given next.

1. Choose the **Hole** tool from the **Feature** group; the **Hole** dialog box is displayed. Next, you need to specify a point to define the hole location.

2. Move the cursor over the lower right round edge of the base feature and specify a point when the cursor snaps to its center point.

3. Select the **Counterbored** option from the **Form** drop-down list in the **Form and Dimensions** rollout of the dialog box. Make sure **General Hole** is selected in the type drop-down list of the dialog box.

4. Enter **20** in the **C-Bore Diameter** edit box, **2** in the **C-Bore Depth** edit box, and **10** in the **Diameter** edit box of the **Dimensions** sub-rollout of the **Form and Dimensions** rollout.

5. Select the **Until Selected** option from the **Depth Limit** drop-down list of the **Dimensions** sub-rollout in the dialog box and select the back face of the base feature to define the through face.

6. Choose the **OK** button from the dialog box; the counterbore hole is created, as shown in Figure 6-34.

7. Similarly, create the other three counterbore holes in the base feature of the model. The model after creating all counterbore holes is shown in Figure 6-35.

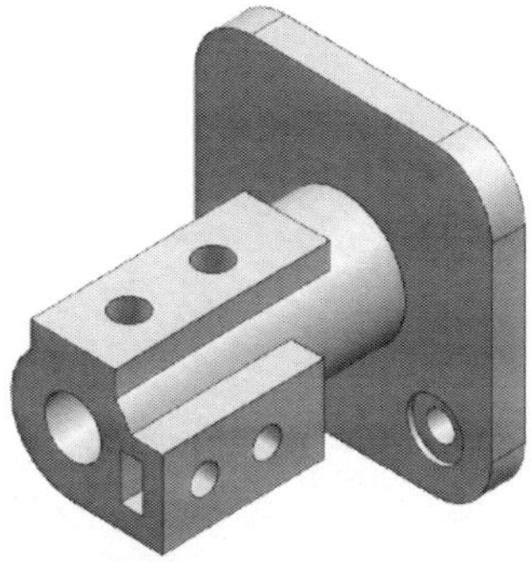

Figure 6-34 *The model after creating one counterbore hole*

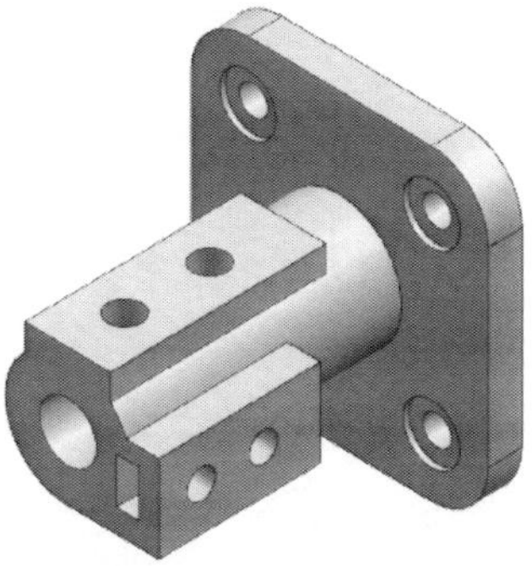

Figure 6-35 *The model after creating all counterbore holes*

Saving and Closing the File

1. Choose **Menu > File > Close > Save and Close** from the **Top border Bar** to save and close the current file.

EXERCISES

Exercise 1

Create the model shown in Figure 6-36. The drawing views and dimensions of the model are given in Figure 6-37. After creating the model, save it with the name *c06exr1.prt* at the following location: *\NX\c06*. **(Expected time: 30 min)**

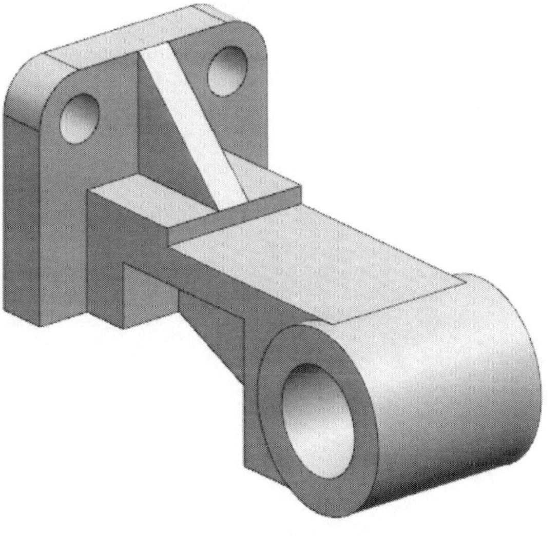

Figure 6-36 *Model for Exercise 1*

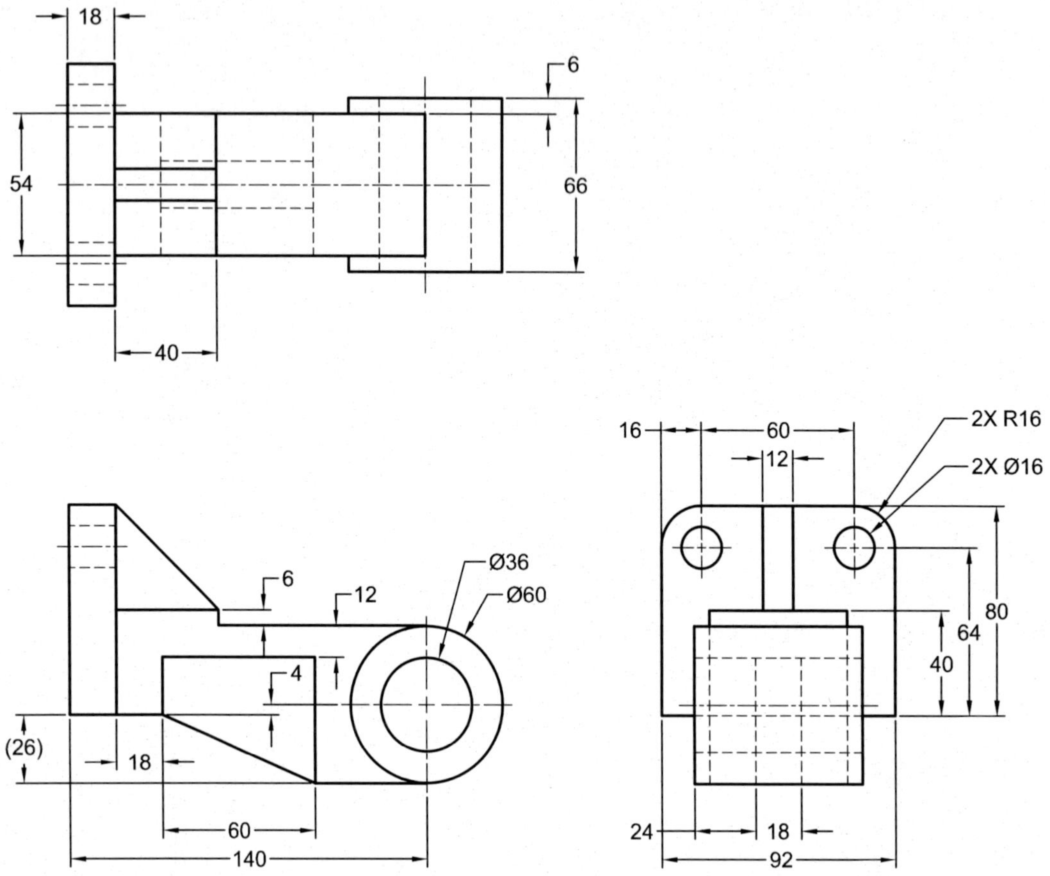

Figure 6-37 Drawing views and dimensions of the model for Exercise 1

Exercise 2

Create the model shown in Figure 6-38. The drawing views and dimensions of the model are given in Figure 6-39. After creating the model, save it with the name *c06exr2.prt* at the following location: *\NX\c06*.

(Expected time: 30 min)

Figure 6-38 Model for Exercise 2

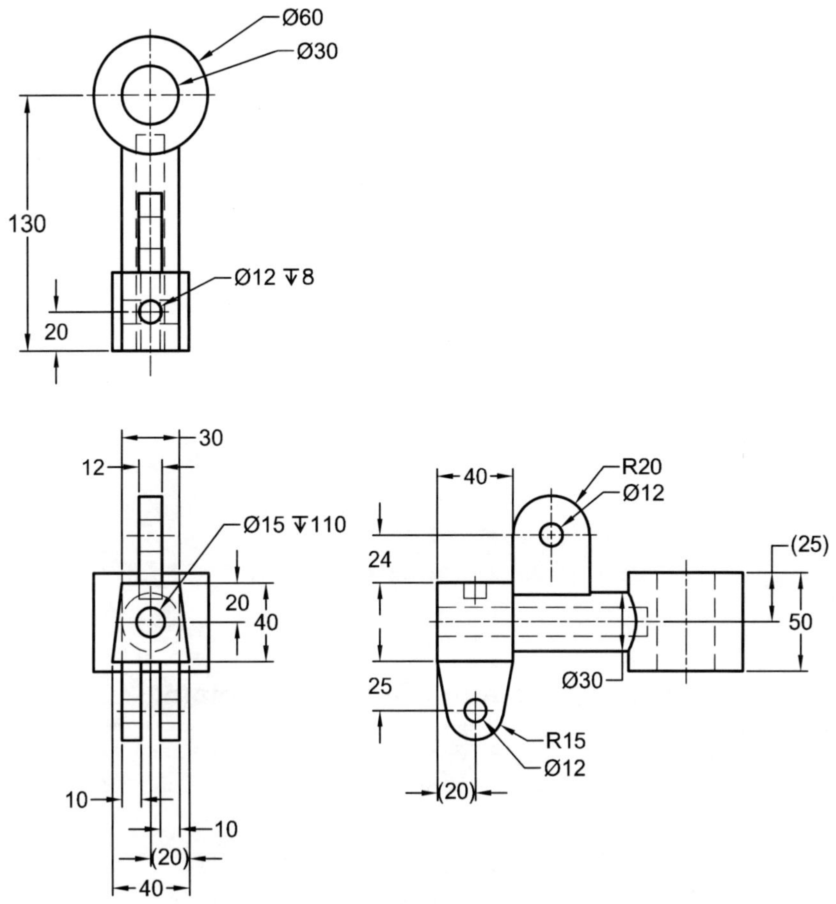

Figure 6-39 *Drawing views and dimensions of the model for Exercise 2*

This page is intentionally left blank

Chapter 7

Advanced Modeling Tools-II

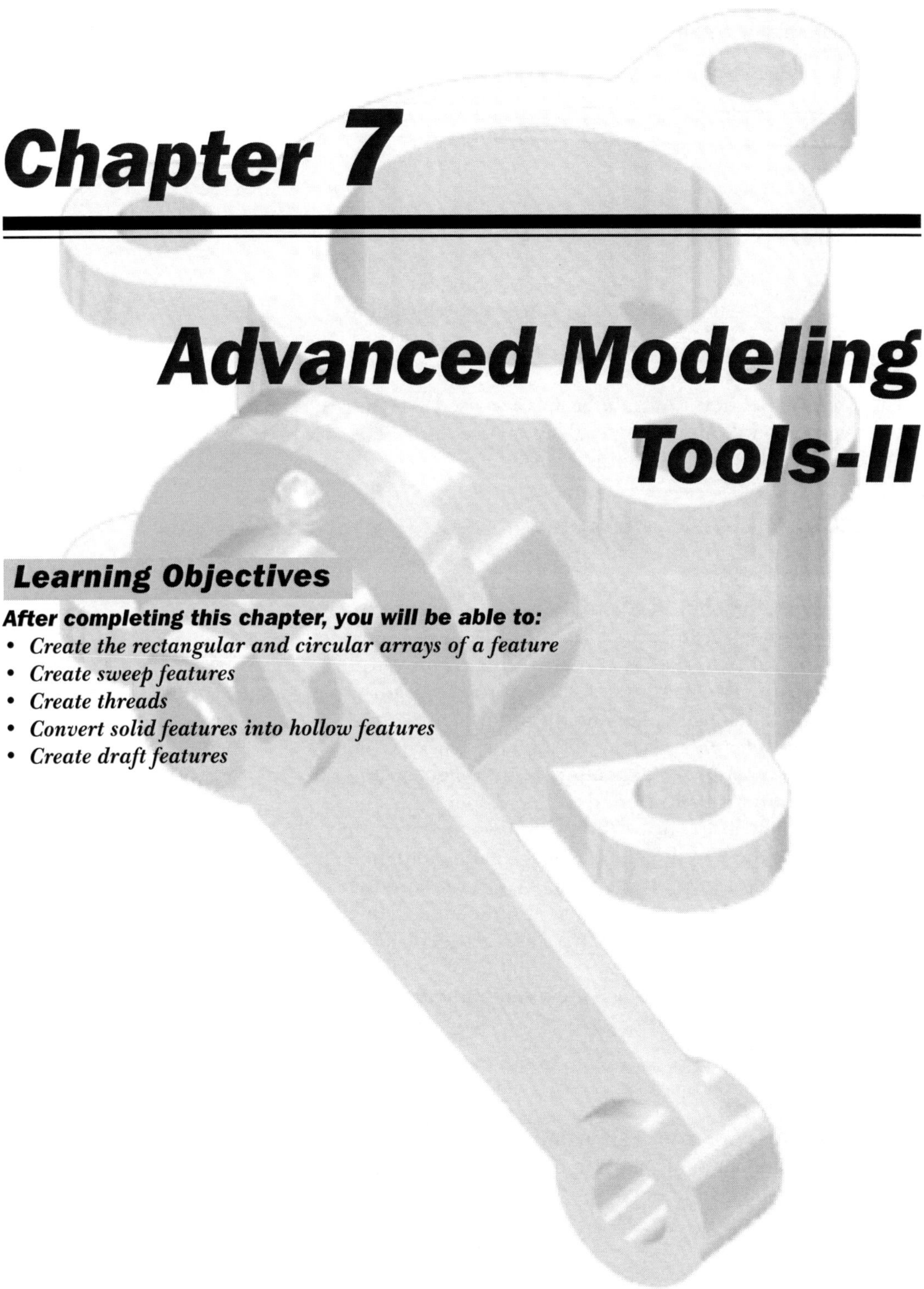

Learning Objectives

After completing this chapter, you will be able to:

- *Create the rectangular and circular arrays of a feature*
- *Create sweep features*
- *Create threads*
- *Convert solid features into hollow features*
- *Create draft features*

ADVANCED MODELING TOOLS

In this chapter, you will learn about the following advanced modeling tools:

1. Pattern Feature
2. Mirror Feature
3. Mirror Body
4. Sweep along Guide
5. Swept
6. Tube
7. Thread
8. Shell

PATTERN FEATURE TOOL

Ribbon:	Home > Feature > Pattern Feature
Menu:	Insert > Associative Copy > Pattern Feature

The **Pattern Feature** tool is a very versatile tool in NX. This tool is used to create the linear and circular patterns of features. It is also used to create the polygon, spiral, along, general, reference, and helix patterns. The uses of this tool are discussed next.

Creating a Linear Pattern

Linear pattern of a feature is created by placing the instances of that feature along the two directions of the work coordinate system (WCS). To create a linear pattern, choose the **Pattern Feature** tool from the **Feature** group of the **Home** tab; the **Pattern Feature** dialog box will be displayed, refer to Figure 7-1. The **Linear** option is selected by default in the **Layout** drop-down list of the **Pattern Definition** rollout of the dialog box. Click in the **Specify Vector** area of the **Direction 1** sub-rollout in the **Pattern Definition** rollout; a triad will be displayed in the graphics window and you will be prompted to select an object to define the first direction. Select an edge to define the first direction of the pattern creation. You can also use the triad to define the direction. Next, you need to define the number of instances and the spacing between them. There are four options to define the number of instances and spacing between them, which are listed in the **Spacing** drop-down list in the **Direction 1** sub-rollout. These options are discussed next.

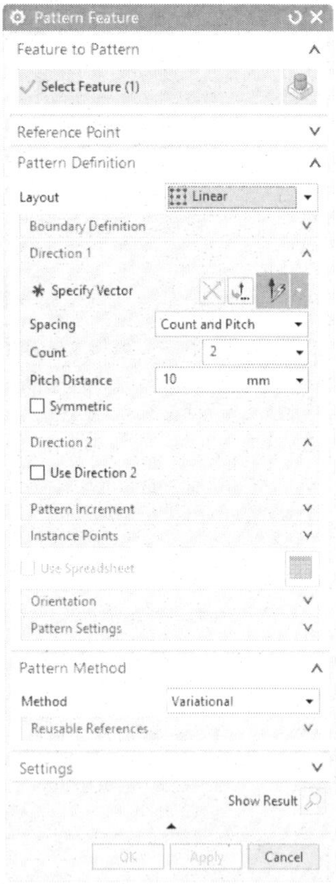

Figure 7-1 *The **Pattern Feature** dialog box for creating a linear pattern*

Count and Pitch

On selecting this option, the **Count** and **Pitch Distance** edit boxes will be displayed. You need to specify the number of instances in the **Count** edit box and the distance between the instances in the **Pitch Distance** edit box.

Count and Span

On selecting this option, the **Count** and **Span Distance** edit boxes will be displayed. You need to specify the number of instances in the **Count** edit box and the total length of the pattern in the **Span Distance** edit box.

Pitch and Span

On selecting this option, the **Pitch Distance** and **Span Distance** edit boxes will be displayed. You need to specify the distance between two consecutive instances in the **Pitch Distance** edit box and the total length of the pattern in the **Span Distance** edit box.

List

On selecting this option, a table will be displayed in the **Direction1** sub-rollout. Also, the **Spacing Value** edit box will be displayed. You need to enter a value in the **Spacing Value** edit box. You can choose the **Add New Set** button located next to the table to add a new spacing value. You can remove a spacing value from the table using the **Remove** button.

After specifying the count and spacing of the pattern, you can select the **Symmetric** check box to create the pattern on both sides of the feature. Figure 7-2 shows the preview of the pattern created on both sides of the feature.

You can also create a linear pattern in two directions. To do so, select the **Use Direction 2** check box in the **Direction 2** sub-rollout; the options in this sub-rollout will be displayed. Next, specify the parameters in the **Direction 2** sub-rollout. The options in this sub-rollout are same as in the **Direction 1** sub-rollout. Figure 7-3 shows a model with a linear pattern created in two directions.

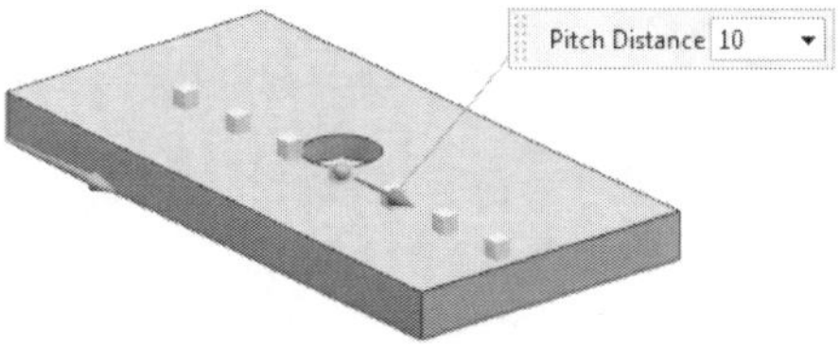

Figure 7-2 *Preview of the pattern created on both sides of the feature*

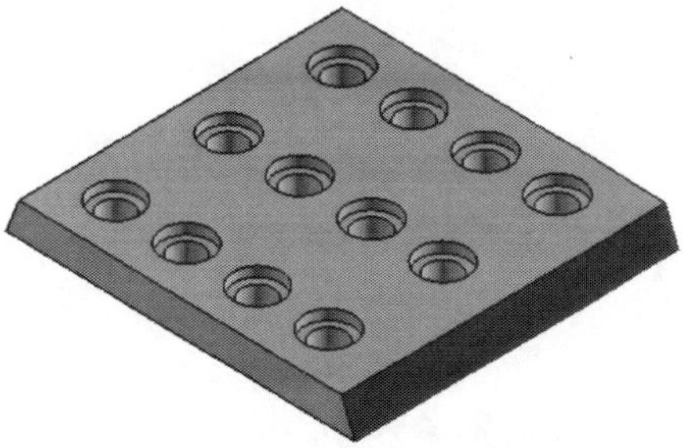

Figure 7-3 *Linear pattern created in two directions*

The advanced options of the **Pattern Feature** dialog box are discussed next.

Creating a Circular Pattern

To create a circular pattern of the selected features, invoke the **Pattern Feature** tool; the **Pattern Feature** dialog box will be displayed. Select the **Circular** option from the **Layout** drop-down list in this dialog box; the **Pattern Definition** rollout in this dialog box will be modified, as shown in Figure 7-4, and you will be prompted to select features to pattern. Select the features to be patterned from the drawing window. Next, click in the **Specify Vector** area in the **Rotation Axis** sub-rollout of the **Pattern Definition** rollout; a triad will be displayed in the drawing window and you will be prompted to select an object to infer vector. Use the triad to specify the axis of rotation. Note that you need to select the axis perpendicular to the face on which the pattern is to be created. Next, you need to specify the center of the circular pattern. You can infer to center point of the circular face to specify the center of the pattern. Next, you need to define the number of instances and the angular spacing between them. There are four options to define the number of instances and spacing between them. These options are listed in the **Spacing** drop-down list in the **Angular Direction** sub-rollout and they are discussed next.

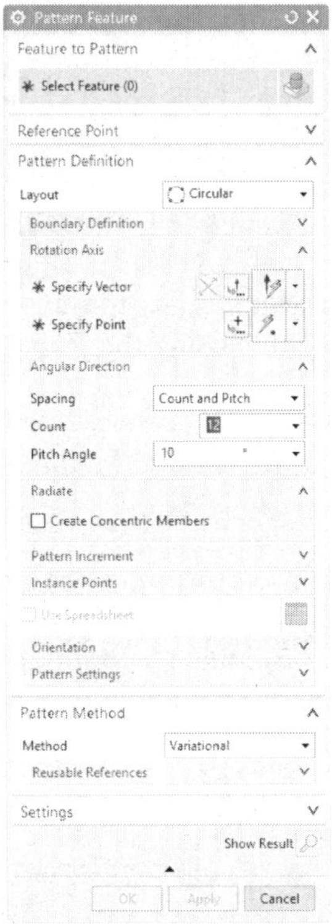

*Figure 7-4 The **Pattern Definition** rollout for creating a circular pattern*

Count and Pitch

This option is used to specify the instance count and the angle between the instances.

Count and Span

This option is used to specify the instance count and the span angle of the pattern.

Pitch and Span

This option is used to specify the angle between the instances and the span angle of the pattern.

List

This option is used to specify separate angle between individual instances of the pattern. On selecting this option, a table will be displayed in the **Angular Direction** sub-rollout. Also, the

Spacing Value edit box will be displayed in which you need to enter an angular value. You can choose the **Add New Set** button located next to the table to add a new spacing value. You can remove a spacing value from the table using the **Remove** button.

SWEEPING SKETCHES ALONG THE GUIDE CURVES

Ribbon:	Home > Feature > More Gallery > Sweep > Sweep along Guide *(Customize to Add)*
Menu:	Insert > Sweep > Sweep along Guide

The **Sweep along Guide** tool is used to create sweep features by sweeping an open or a closed section about a guide string. Figure 7-5 shows an open section and the guide string and Figure 7-6 shows the resulting sweep feature.

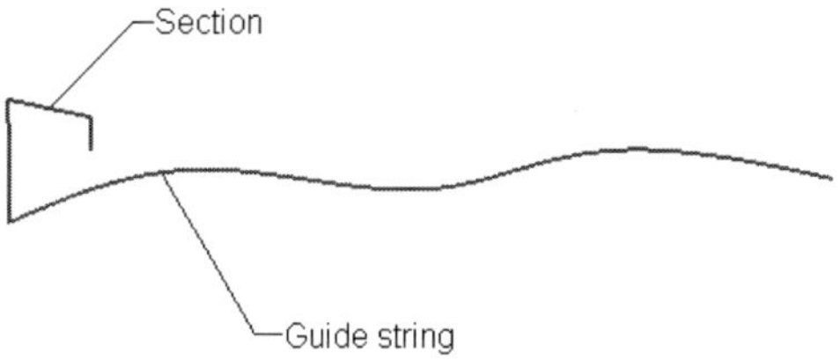

Figure 7-5 *An open section and a guide string to create a sweep feature*

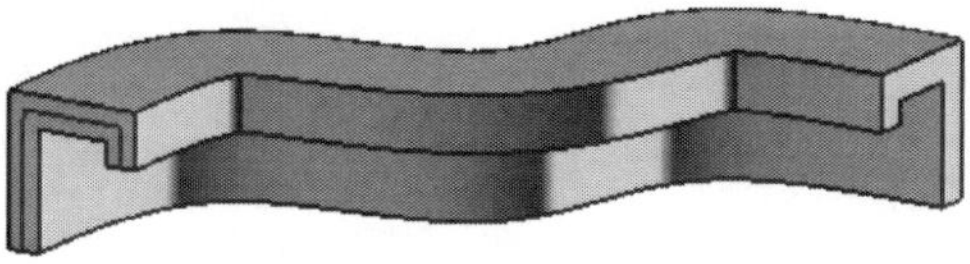

Figure 7-6 *The resulting sweep feature*

To invoke this tool, choose **Menu** > **Insert > Sweep > Sweep along Guide** from the **Top Border Bar**; the **Sweep Along Guide** dialog box will be displayed, as shown in Figure 7-7, and you will be prompted to select the chain of curves for the section. Select the chain of curves from the drawing window. Next, choose the **Curve** button from the **Guide** rollout; you will be prompted to select the chain of curves for the guide. Note that if the guide has multiple strings, they all should be end-connected to each other. After selecting the guide curve, enter values for the first and second directions in the **First Offset** and **Second Offset** edit boxes of the **Offsets** rollout, respectively. Note that these values should not be the same; it means one value must be greater than the other. Also, an arrow will be displayed in the graphics window, pointing toward the second offset direction. You can use this arrow to dynamically increase or decrease the values of the first and second offset directions.

Select the required boolean operation from the **Boolean** drop-down list in the dialog box. Note that if the sweep feature is the base feature in the model, then only the **None** option will be available in the **Boolean** drop-down list. Next, choose the **OK** button from the **Sweep Along Guide** dialog box; the sweep feature will be created.

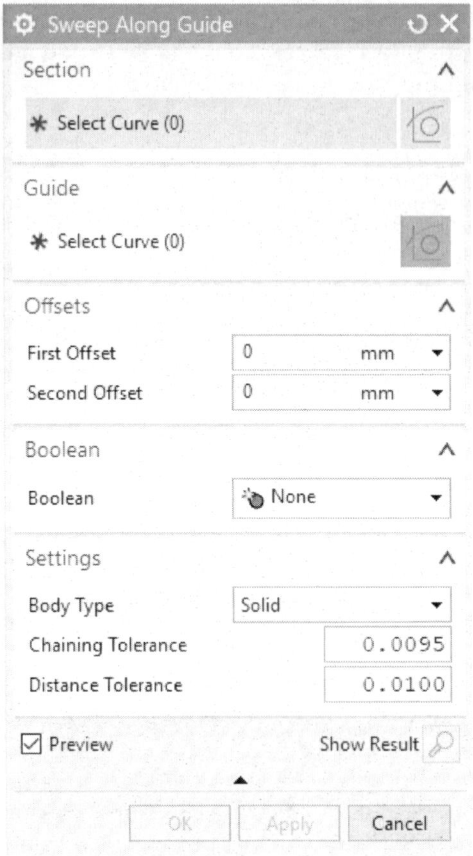

*Figure 7-7 The **Sweep Along Guide** dialog box*

CREATING THREADS

Ribbon: Home > Feature > More Gallery > Design Feature > Thread
Menu: Insert > Design Feature > Thread

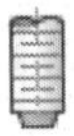

NX allows you to create internal or external threads on a cylindrical model. Internal threads are those that are created on the internal faces of a model, similar to a hole in a model. External threads are those that are created on the outer face of a model. Figure 7-8 shows the external threading on a bolt and Figure 7-9 shows the internal threading in a nut.

Figure 7-8 *External threading on a bolt*

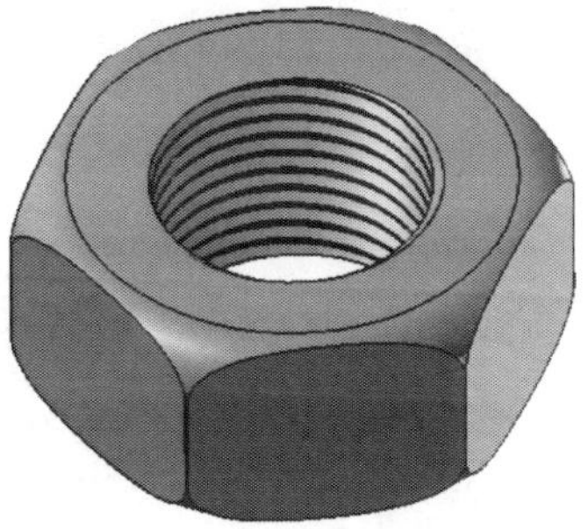

Figure 7-9 *Internal threading in a nut*

In NX, you can create detailed or symbolic threads. The detailed threads appear like actual threads, refer to Figures 7-8 and 7-9 whereas symbolic threads just display the thread convention in the form of dashed lines, as shown in Figure 7-10.

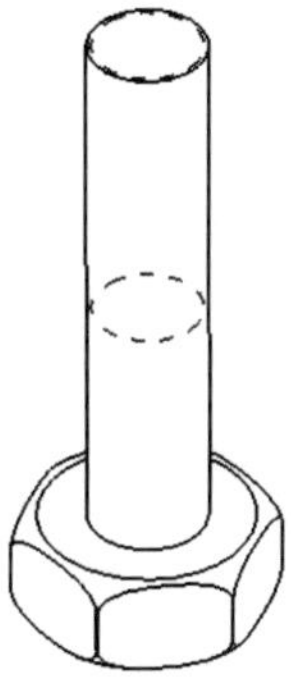

Figure 7-10 *Symbolic thread on a bolt*

Creating Symbolic Threads

To create symbolic threads, invoke the **Thread** tool; the **Thread** dialog box will be displayed. Select the **Symbolic** radio button, if it is not selected by default, to display the options to create symbolic threads, refer to Figure 7-11.

There are two methods of creating symbolic threads. In the first method, the system automatically defines the thread specifications for the selected face using a thread table. In the other method, some thread specifications are calculated and displayed in the edit box, but you can also modify these values individually. Both these methods are discussed next.

Creating Symbolic Threads Using Table

To create a symbolic thread using table, invoke the **Thread** dialog box and clear the **Manual Input** check box if it is already selected; the options to specify the parameters of the thread will be disabled. Next, select the cylindrical face on which you need to create the threads; the face will be highlighted and an arrow will be displayed at one of the ends of the face. The options that will be available in the dialog box at this stage are discussed next.

Shaft Size / Tapped Drill Size

If you select an outer face to create threads, the **Shaft Size** edit box will be available. In this edit box, you can specify the size of the shaft. But if you select an interior face to create the thread, the **Tapped Drill Size** edit box will be available. Depending upon the face selected and the specifications from the table, a default value will be displayed in this edit box. You can accept the default value or enter a new value.

Method

This drop-down list is used to select the method that will be used to create threads. You can select the **Cut**, **Rolled**, **Ground**, or **Milled** option from this drop-down list.

Pitch
This edit box is used to specify the pitch of threads.

Angle
This edit box is used to specify the angle of threads.

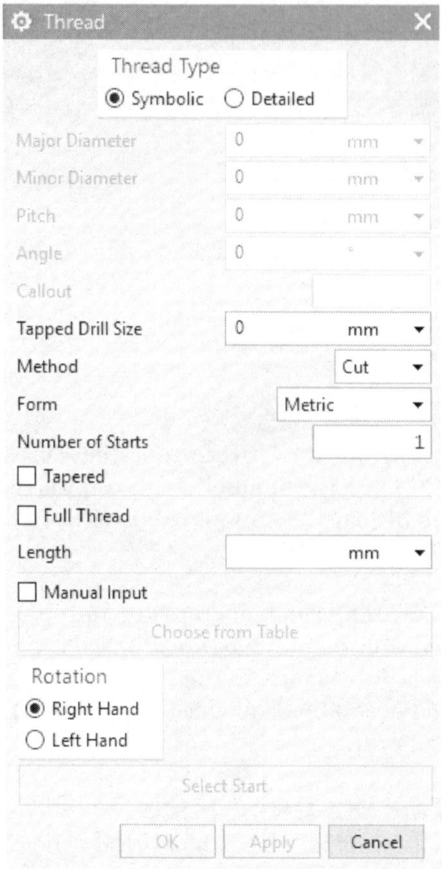

***Figure 7-11** The **Thread** dialog box*

Creating Detailed Threads
To create the detailed thread, invoke the **Thread** tool; the **Thread** dialog box will be displayed. Select the **Detailed** radio button to display the options to create detailed threads; the dialog box will be modified and you will be prompted to select a cylindrical face, refer to Figure 7-12. Select the face on which you need to create threads; the options in the dialog box will be enabled. Most of the options in this dialog box are similar to those discussed while creating symbolic threads. The **Length** option is discussed next.

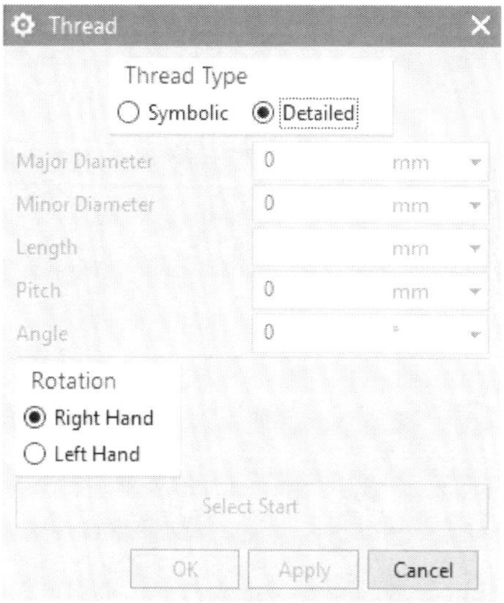

***Figure 7-12** The **Thread** dialog box*

Length

This edit box is used to specify the depth up to which the threads will be created. The default value that is displayed in this edit box is equal to the length of the cylindrical face that is selected to create threads. You can enter any other value in this edit box. Figure 7-13 shows a bolt with threads created throughout its length and Figure 7-14 shows a bolt with threads created up to a specified length.

***Figure 7-13** Threads created throughout the length of the bolt*

***Figure 7-14** Threads created up to a specified length*

CREATING SHELL FEATURES

Ribbon:	Home > Feature > Shell
Menu:	Insert > Offset/Scale > Shell

The **Shell** tool allows you to scoop out the material from a model and make it hollow. The resultant model will be a structure of walls with a cavity inside. You can remove some of the faces of the model and apply different wall thicknesses to them. Figure 7-15 shows a solid model before scooping out material and Figure 7-16 shows the model after shelling it.

Figure 7-15 *Solid model before scooping out material*

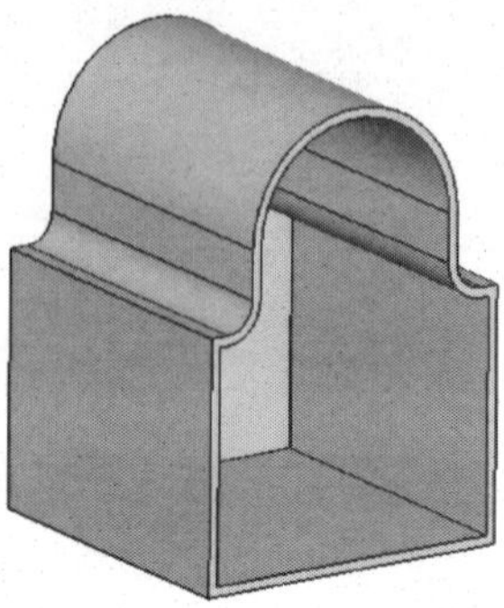

Figure 7-16 *Solid model after scooping out material using the **Shell** tool*

On invoking this tool, the **Shell** dialog box will be displayed, refer to Figure 7-17. You can hollow the models using various rollouts of this dialog box.

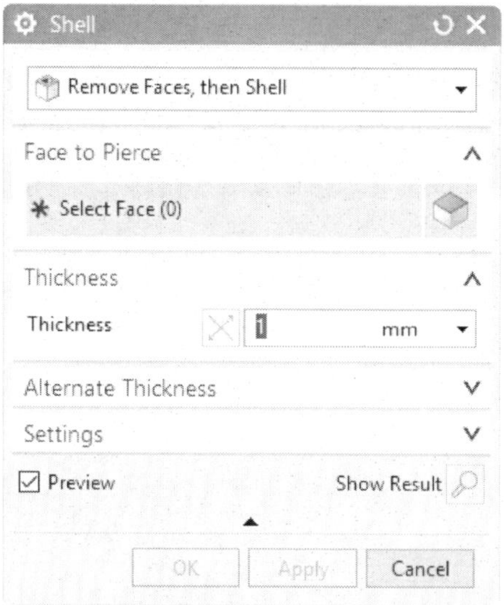

Figure 7-17 *The **Shell** dialog box*

Creating Drafts

Ribbon:	Home > Feature > Draft
Menu:	Insert > Detail Feature > Draft

This tool is used to create a draft on an existing model. To create the draft, choose the **Draft** tool from the **Feature** group of the **Home** tab; the **Draft** dialog box will be displayed, as shown in Figure 7-18. In NX, you can create four types of drafts: draft from face, draft from edges, draft tangent to faces, and draft to parting edges. The Type drop-down list contains four options to create the draft.

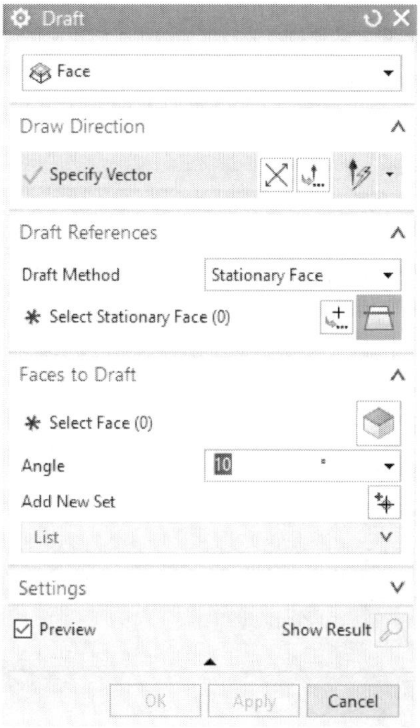

Figure 7-18 *The **Draft** dialog box*

TUTORIALS

Tutorial 1

In this tutorial, you will create the model shown in Figure 7-19. The dimensions of the model are shown in Figure 7-20. **(Expected time: 30 min)**

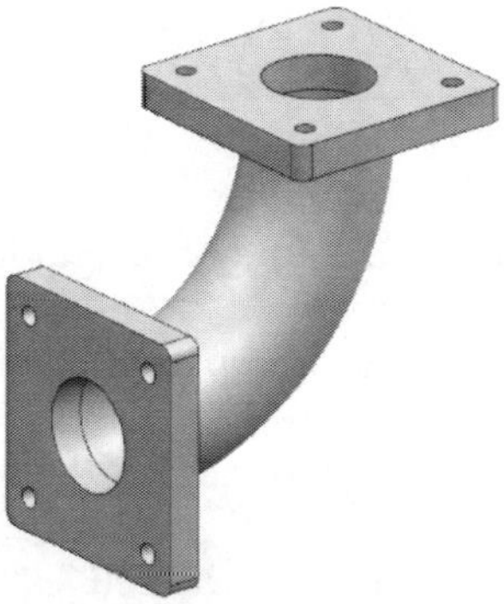

Figure 7-19 *Model for Tutorial 1*

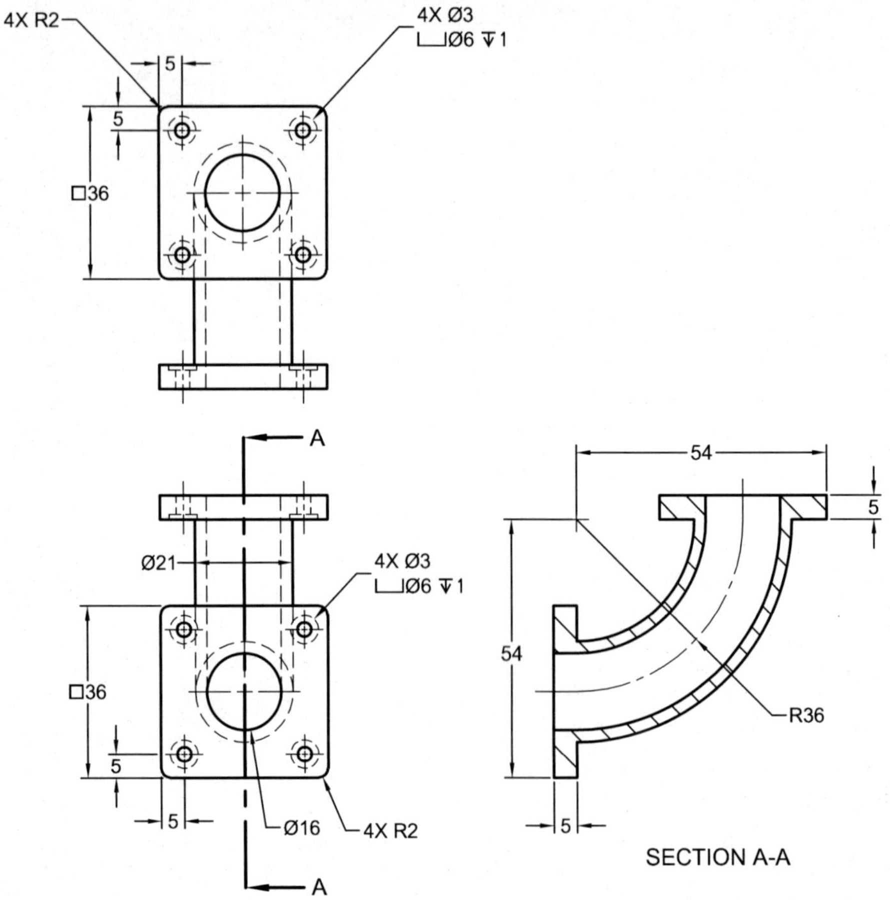

Figure 7-20 *Views and dimensions of the model*

The following steps are required to complete this tutorial:

a. Draw the sketch of the guide curve on the YC-ZC plane for the base feature, refer to Figure 7-21.
b. Create a new datum plane at the start point of the guide curve and then draw the sketch of the section on this plane, refer to Figure 7-22.
c. Create the sweep feature by using the **Sweep along Guide** tool, refer to Figure 7-23.
d. Hollow the sweep feature, refer to Figure 7-24.
e. Create extruded join features on the end faces of the sweep feature and then create holes on them, refer to Figures 7-25 and 7-26.
f. Create the counterbore hole on one of the extruded features and then create the rectangular pattern of holes, refer to Figure 7-27 and 7-28.
g. Similarly, create the counterbore hole on the other extruded feature and then create the pattern of holes, refer to Figure 7-29.

Creating the Base Feature

The base feature of this model is a sweep feature. First, you will draw the guide curve of the sweep feature and then create a datum plane at the lower end of the guide curve. You will use this datum plane as the sketching plane to sketch the section of the sweep feature.

1. Start a new file with the name *c07tut1.prt* using the **Model** template and specify its location as *C:\NX\c07*.
2. Draw the sketch of the guide curve on the Y-Z plane, refer to Figure 7-21.
3. Create a datum plane normal to the curve at its lower end and then draw the sketch of the section on this plane, refer to Figure 7-22.

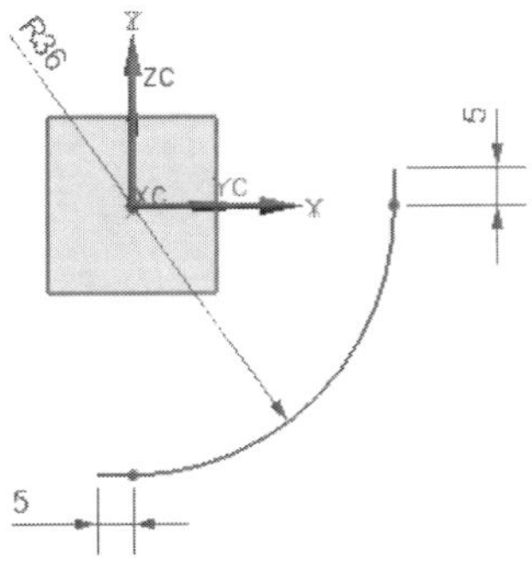

Figure 7-21 *Sketch of the guide curve*

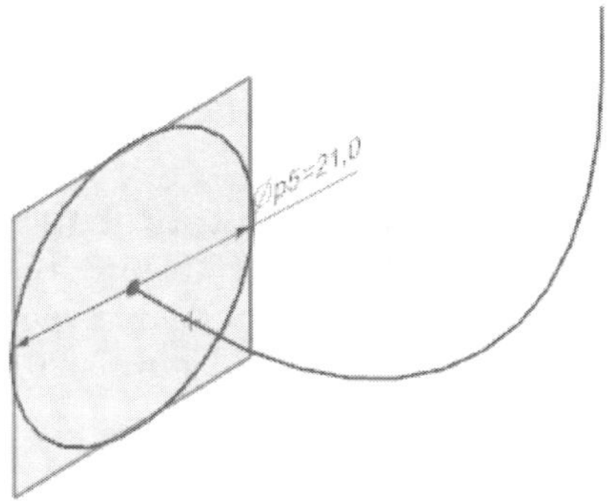

Figure 7-22 *Sketch of the section*

4. After creating the guide curve and a section of the sweep feature, choose **Menu > Insert > Sweep > Sweep along Guide** from the **Top Border Bar**; the **Sweep Along Guide** dialog box is displayed and you are prompted to select the section string.
5. Select the circle from the drawing area and then press the middle mouse button once; you are prompted to select the guide string. Select the guide curve from the drawing area; a preview of the feature is displayed in the drawing window. Make sure the value **0** is entered in the **First Offset** and **Second Offset** edit boxes of the **Offsets** rollout in the dialog box.
6. Choose the **OK** button from this dialog box; the sweep feature is created, as shown in Figure 7-23.

Hollowing out the Base Feature

Next, you need to hollow out the base feature by using the **Shell** tool.

1. Invoke the **Shell** tool from the **Feature** group; the **Shell** dialog box is displayed and you are prompted to select the faces to be removed.

2. Enter **2.5** in the **Thickness** edit box and select the two end faces of the sweep feature.
3. Choose **Apply** and then the **Cancel** button to exit the dialog box. The model after hollowing out the base feature is shown in Figure 7-24.

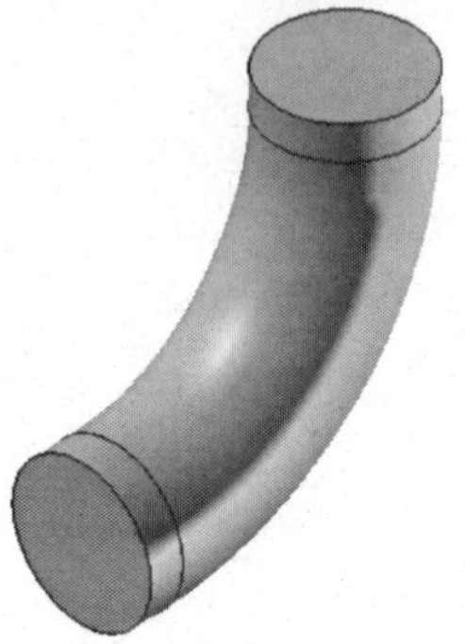

Figure 7-23 *Base feature of the model*

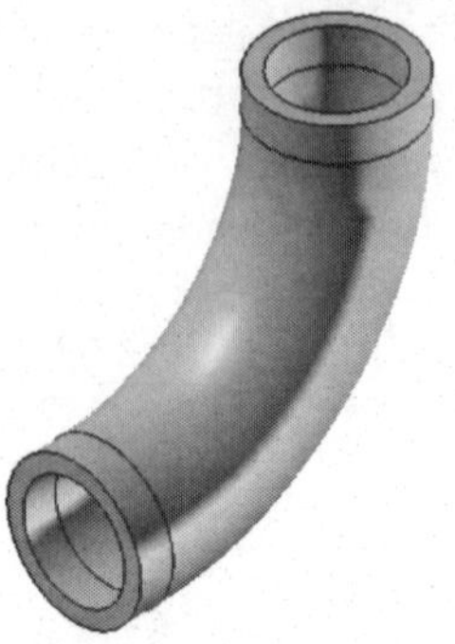

Figure 7-24 *Model after hollowing the base feature*

Note

For hollowing the base feature, you can also enter ***-2.5*** *in the* ***First Offset*** *edit box of the* ***Offsets*** *rollout in the* ***Sweep Along Guide*** *dialog box while creating the base feature.*

Creating Extruded Join Features

1. Create the extruded join features one by one on the two end faces of the base feature, as shown in Figure 7-25. Note that you need to reverse the direction of the feature creation. For dimensions, refer to Figure 7-20.

2. Create simple holes at the center of the extruded join features. Note that the holes need to be created only up to the back faces of the extruded features. For diameters of the holes, refer to Figure 7-20. Figure 7-26 shows the model after the holes have been created.

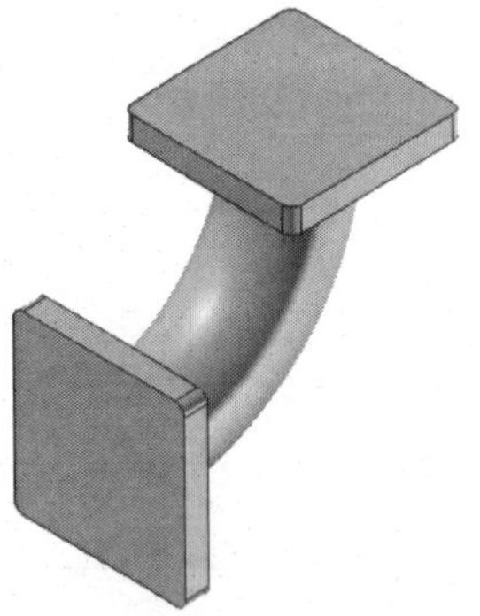

Figure 7-25 *Extruded join features created on the two end faces of the base feature*

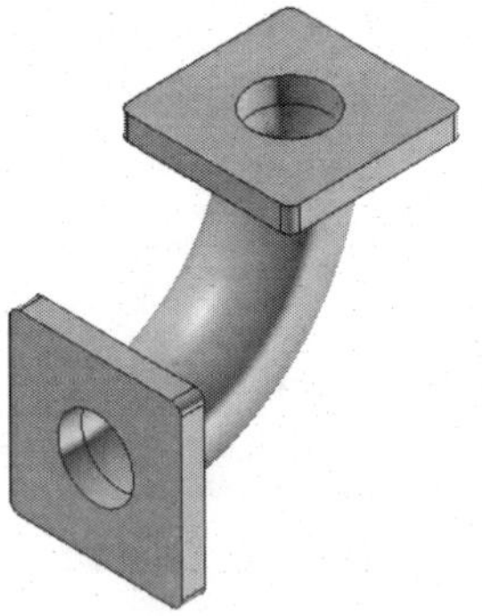

Figure 7-26 *Model after creating holes*

Creating Counterbore Holes

1. Create the counterbore hole on the lower right of the back face of the front extruded feature of the model. The rotated view of the model after creating the hole is shown in Figure 7-27. For dimensions of the counterbore hole, refer to Figure 7-20.

2. After creating the counterbore hole, choose the **Pattern Feature** tool from the **Feature** group; the **Pattern Feature** dialog box is displayed and you are prompted to select the features to be patterned.

3. Select the **Linear** option from the **Layout** drop-down list if it is not already selected; the options related to the linear pattern are displayed.

4. Select the counterbore hole from the drawing window and choose the **Specify Vector** option from the **Direction 1** sub-rollout in the **Pattern Feature** dialog box; the vector triad is displayed in the drawing window and you are prompted to select objects to infer vector.

5. Select the lower horizontal edge of the extruded join feature to define the first direction. Next, you need to define the second direction.

6. Select the **Use Direction 2** check box from the **Direction 2** sub-rollout; the options in the **Direction 2** sub-rollout are displayed and you are prompted to select objects to infer vector.

7. Select the vertical edge of the extruded join feature to define the second direction.

   Next, you need to define the number of instances and the spacing between them.

Note

If the direction arrows point in the directions opposite to what is required, then choose the ***Reverse Direction*** *button.*

8. Select the **Count and Pitch** option from the **Spacing** drop-down list in the **Direction 1** sub -rollout; the **Count** and **Pitch Distance** edit boxes are displayed.

9. Enter **2** in the **Count** edit box and **26** in the **Pitch Distance** edit box.

   Next, you need to define the parameters in the second direction.

10. Specify the same parameters in the **Direction 2** sub-rollout and choose the **OK** button from the **Pattern Feature** dialog box; a linear pattern of counterbore holes is created. The rotated view of the model after creating the linear pattern of the counterbore holes is shown in Figure 7-28.

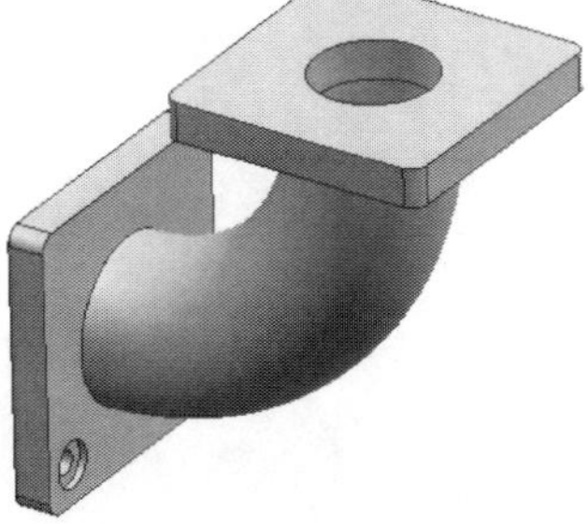

Figure 7-27 *One of the counterbore holes created on the front face of the model*

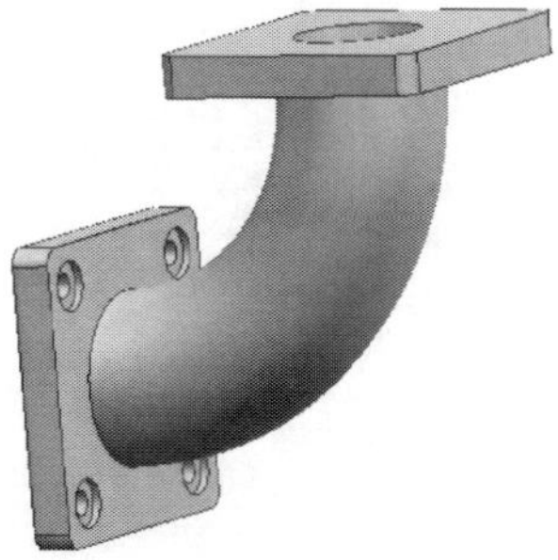

Figure 7-28 *Model after creating the linear pattern of counterbore holes*

11. Similarly, create the counterbore hole on the bottom face of the top extruded feature and then create a linear pattern of counterbore holes on the bottom face of the top extruded feature. The final model for this tutorial is shown in Figure 7-29.

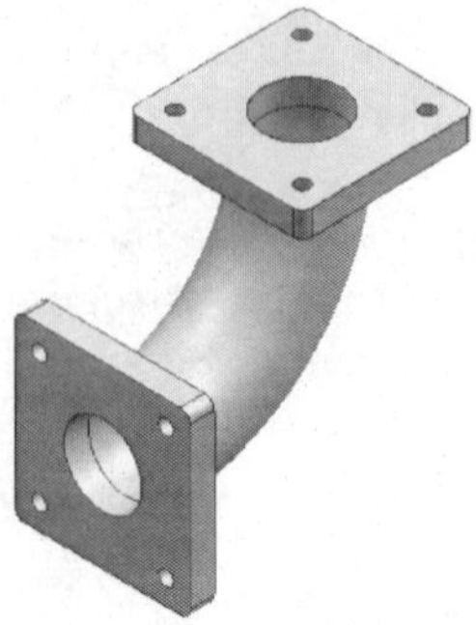

Figure 7-29 Final model for Tutorial 3

Saving and Closing the File

1. Choose **Menu > File > Close > Save and Close** from the **Top Border Bar** to save and close the file.

Tutorial 2

In this tutorial, you will create the model shown in Figure 7-30. The views and dimensions of the model are given in Figure 7-31. **(Expected time: 45 min)**

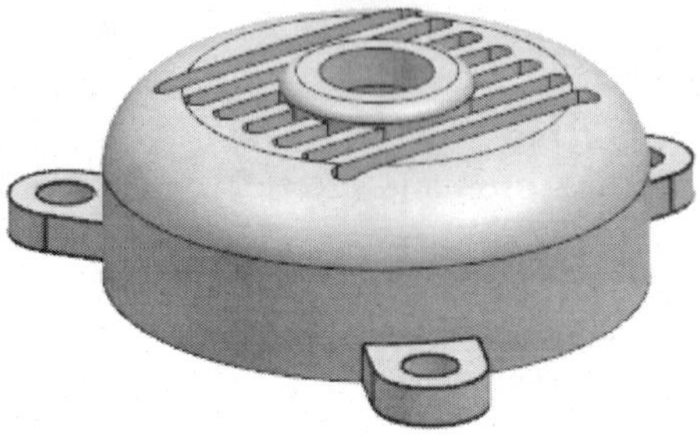

Figure 7-30 Model for Tutorial 2

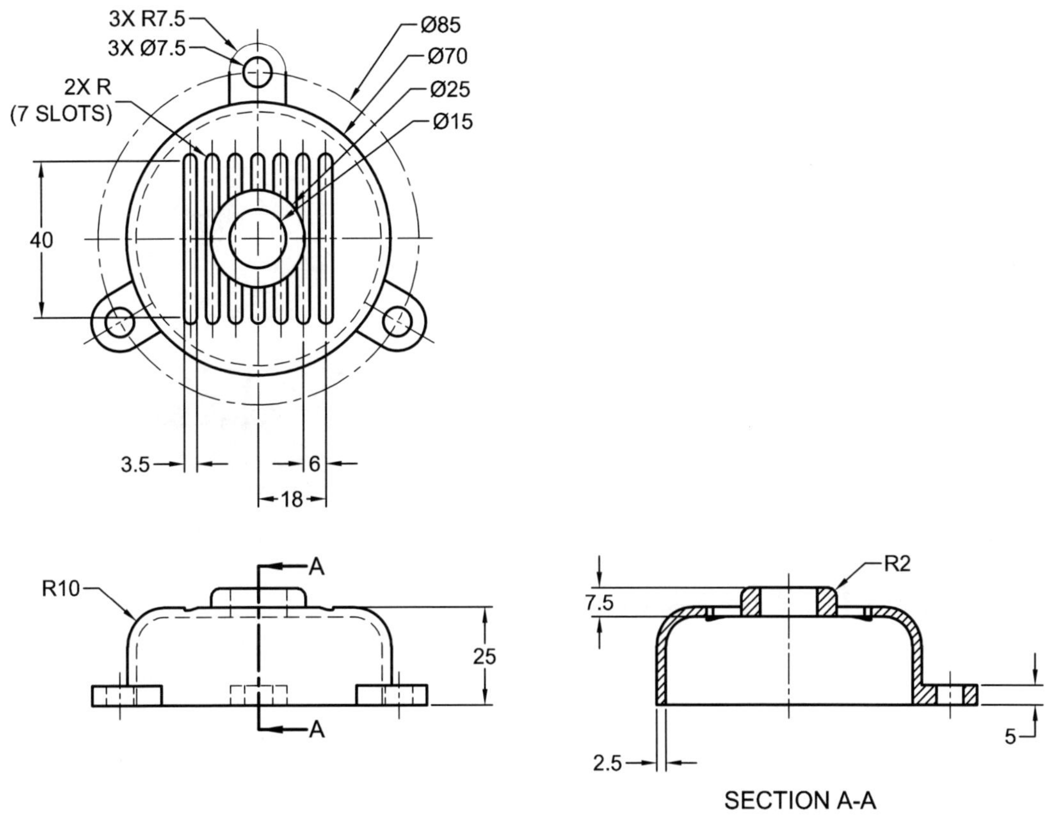

Figure 7-31 Views and dimensions of the model for Tutorial 2

The following steps are required to complete this tutorial:

a. Create the base feature of the model by revolving the sketch drawn on the YC-ZC plane, refer to Figures 7-32 and 7-33.
b. Shell the model using the **Shell** tool, refer to Figure 7-34.
c. Draw the sketch on the XC-YC plane and extrude it to a given distance, refer to Figure 7-35 and 7-36.
d. Pattern the extrude feature using the **Pattern Feature** tool, refer to Figures 7-37.
e. Create the slot on the top planar face and pattern it, refer to Figures 7-38 through 7-41.
f. Create the features on the top planar face, refer to Figures 7-42 and 7-43.
g. Save the model.

Creating the Base Feature

First, you need to create the base feature of the model by revolving the sketch created on the front plane.

1. Start a new file with the name *c07tut2.prt* using the **Model** template and specify its location as *C:\NX\c07*.

2. Invoke the **Revolve** tool and draw the sketch of the base feature on the YC-ZC plane, as shown in Figure 7-32.

3. Exit the Sketch in Task environment and create the base feature of the model, as shown in Figure 7-33.

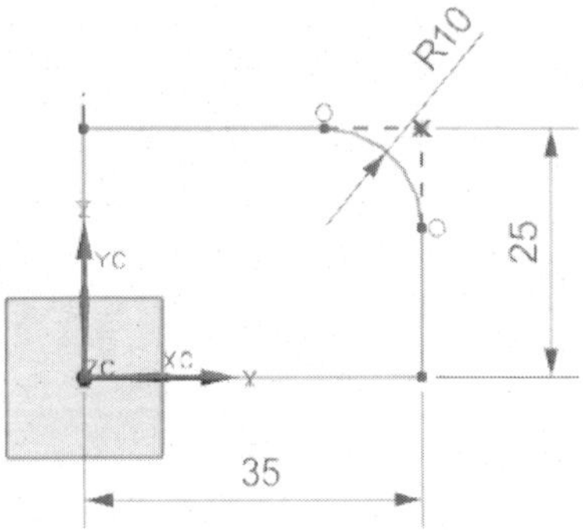

Figure 7-32 *Sketch of the base feature*

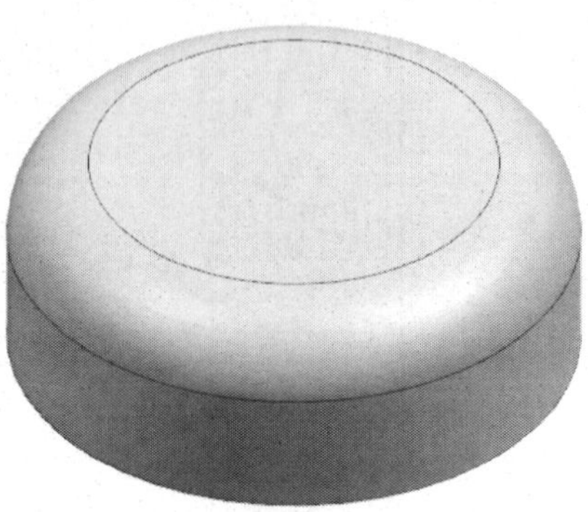

Figure 7-33 *Base feature of the model*

Hollowing out the Base Feature

After creating the base feature, you need to shell the model using the **Shell** tool. You also need to remove the bottom face of the base feature, leaving behind a thin-walled model.

1. Choose the **Shell** tool from the **Feature** group; the **Shell** dialog box is displayed. Set the value in the **Thickness** edit box to **2.5**.

2. Rotate the model and select its bottom face to remove it.

3. Choose the **OK** button to exit the dialog box. The model after hollowing out the base feature is shown in Figure 7-34.

Creating the Third Feature

After adding the shell feature to the model, you need to create the third feature, which is an extruded feature. The sketch for this feature will be drawn on the XC-YC plane.

1. Invoke the **Extrude** tool and select the X-Y plane as the sketching plane.

2. Draw the sketch of the third feature, as shown in Figure 7-35.

3. Exit the Sketch in Task environment and extrude the sketch to a depth of 5 mm.

Figure 7-34 Shell feature added to the model

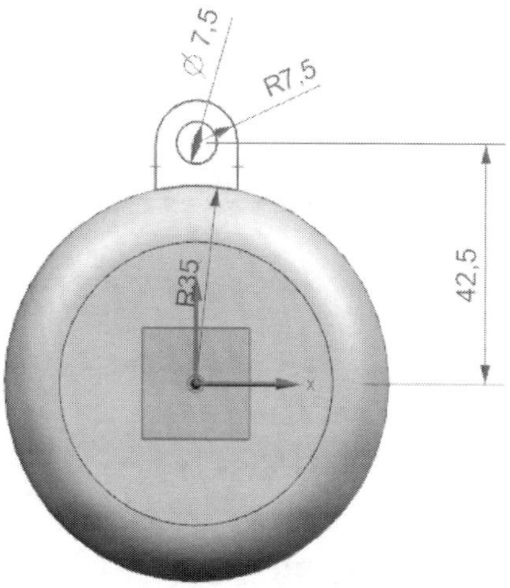

Figure 7-35 Sketch of the third feature

Patterning the Third Feature

You need to pattern the third feature after creating it. This feature will be patterned using the **Pattern Feature** tool.

1. Invoke the **Pattern Feature** dialog box and select the **Circular** option from the **Layout** drop-down list; the options to create a circular pattern are displayed. Next, you need to select the feature to be patterned.

2. Select the third feature from the drawing area.

3. Click in the **Specify Vector** area of the **Rotation axis** sub-rollout; a vector triad is displayed and you are prompted to select object to infer vector. Specify the vector, as shown in Figure 7-36; you are prompted to select object to infer point.

4. Select the center of the circular edge of the base feature; preview of the pattern is displayed. Next, you need to specify the number of instances and spacing between them.

5. Select the **Count and Span** option from the **Spacing** drop-down list.

6. Set the value to **3** in the **Count** edit box and **360** in the **Span Angle** edit box.

7. Select the **Follow Pattern** option from the **Orientation** drop-down list in the **Orientation** sub-rollout and choose **OK** from the **Pattern Feature** dialog box.

The model after creating the pattern feature is displayed in Figure 7-37.

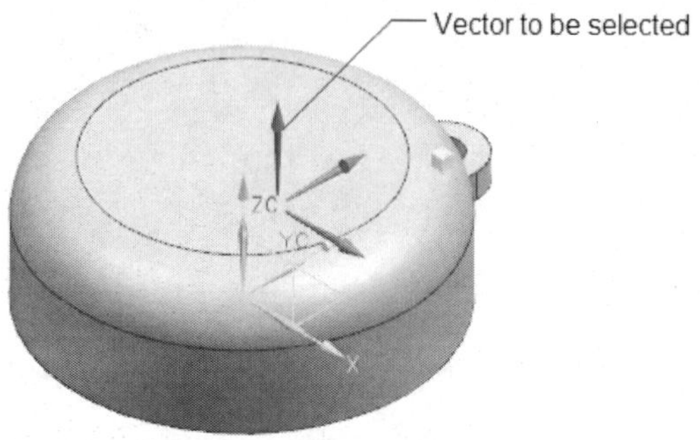

Figure 7-36 Vector to be selected

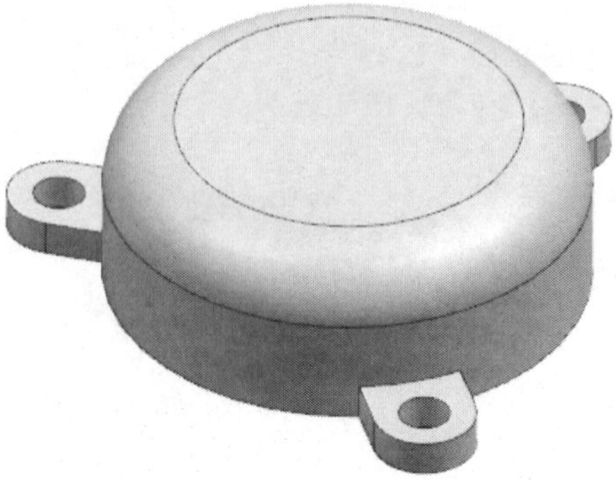

Figure 7-37 Model after creating the pattern

Creating the Slot Cut

Next, you need to create the slots on the top planar face of the base feature.

1. Invoke the **Extrude** tool and select the top planar face of the base feature as the sketching plane.

2. Draw the sketch of the slot and add required dimensions and constraints to it, as shown in Figure 7-38.

3. Exit the Sketch in Task environment and select the **Subtract** option from the **Boolean** drop-down list in the **Extrude** dialog box. On doing so, the preview of the feature is displayed in the drawing window.

4. Choose the **Reverse Direction** button to reverse the direction of the feature creation. Next, select the **Through All** option from the **End** drop-down list in the **Limits** rollout.

5. Choose the **OK** button; the slot feature is created, as shown in Figure 7-39.

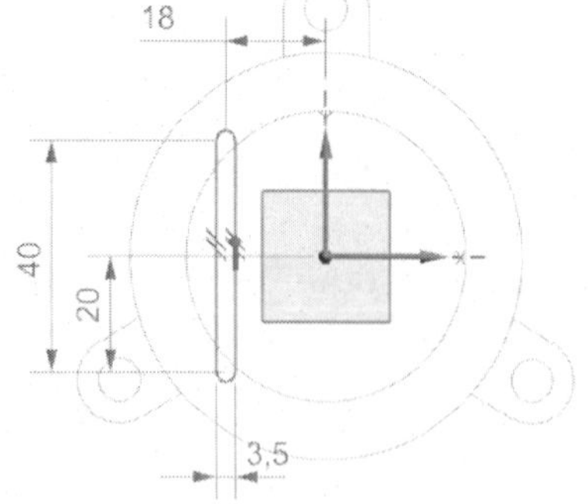

Figure 7-38 Sketch of the slot

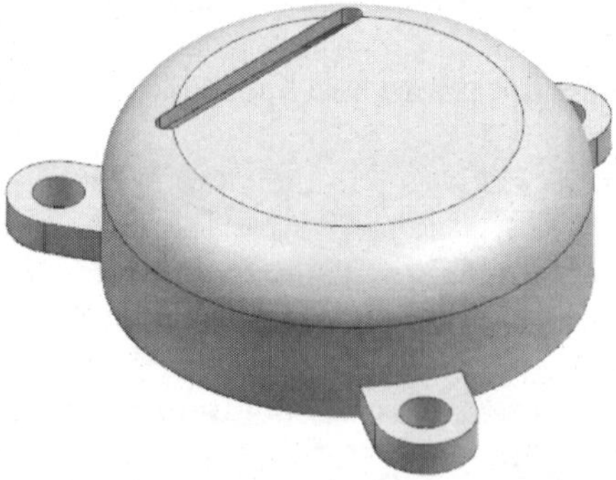

Figure 7-39 Model after creating the slot

6. After creating the slot feature, choose the **Pattern Feature** tool from the **Feature** group; the **Pattern Feature** dialog box is displayed and you are prompted to select features to pattern.

7. Select the **Linear** option from the **Layout** drop-down list; the options related to the linear pattern are displayed.

8. Select the slot from the drawing window and click in the **Specify Vector** area of the **Direction1** sub-rollout in the **Pattern Feature** dialog box; the vector triad is displayed in the drawing window and you are prompted to select objects to infer vector.

9. Select the vector, as shown in Figure 7-40. Next, you need to define the number of instances and spacing between them.

10. Select the **Count and Span** option from the **Spacing** drop-down list in the **Direction1** sub-rollout; the **Count** and **Span Distance** edit boxes are displayed.

11. Enter **7** in the **Count** edit box and **36** in the **Span Distance** edit box and then choose the **OK** button from the **Pattern Feature** dialog box; a linear pattern of slots is created, as shown in Figure 7-41.

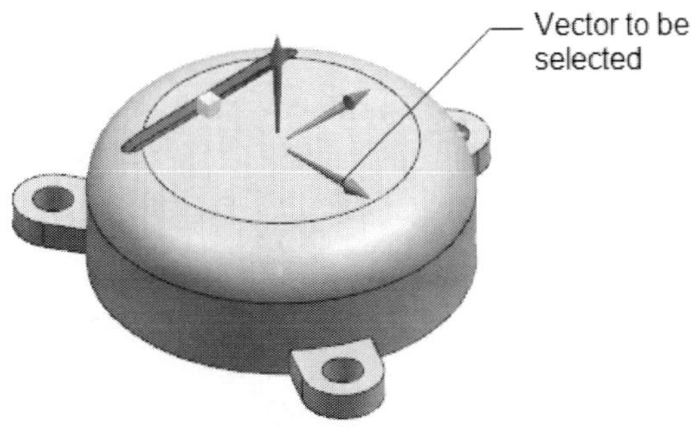

Figure 7-40 *Vector to be selected*

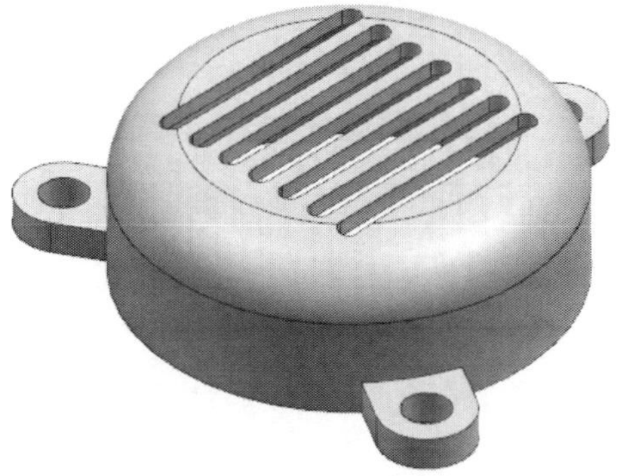

Figure 7-41 *Model after patterning the slot*

Creating the Remaining Features

The next feature that you need to create is the protrusion on the internal face of the base feature. You can create this feature using the **Extrude** tool.

1. Invoke the **Extrude** tool and draw a circle of **25** unit diameter on the internal face of the base feature. Then, extrude it to a distance of **7.5** mm, as shown in Figure 7-42. Choose the **Reverse Direction** button, if required and make sure that the **Unite** option is selected in the **Boolean** drop-down list.

2. Create the remaining features using the **Hole** and **Edge Blend** tools, refer to Figure 7-31 for dimensions. The final model after performing all the operations is shown in Figure 7-43.

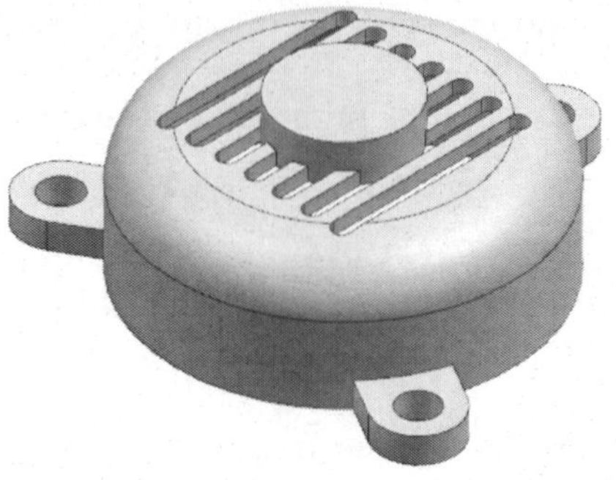

Figure 7-42 The extruded feature

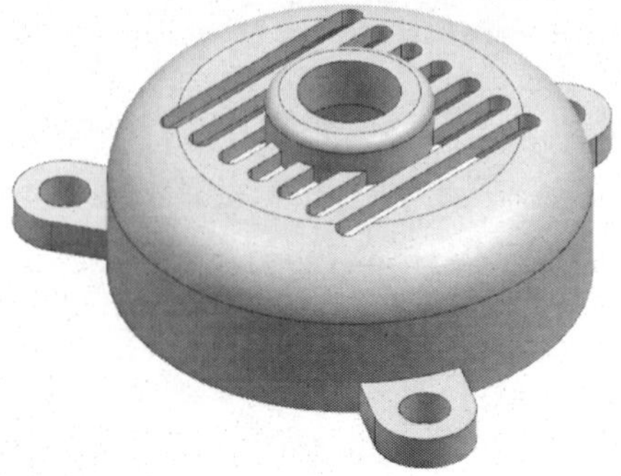

Figure 7-43 Model after creating other features

Saving and Closing the File

1. Choose **Menu > File > Close > Save and Close** from the **Top Border Bar** to save and close the file.

EXERCISES

Exercise 1

Create the model, as shown in Figure 7-44. The dimensions of the model are given in the same figure. Save the file with the name *c07exr1.prt* at the location */NX/c07.*

(Expected time: 30 min)

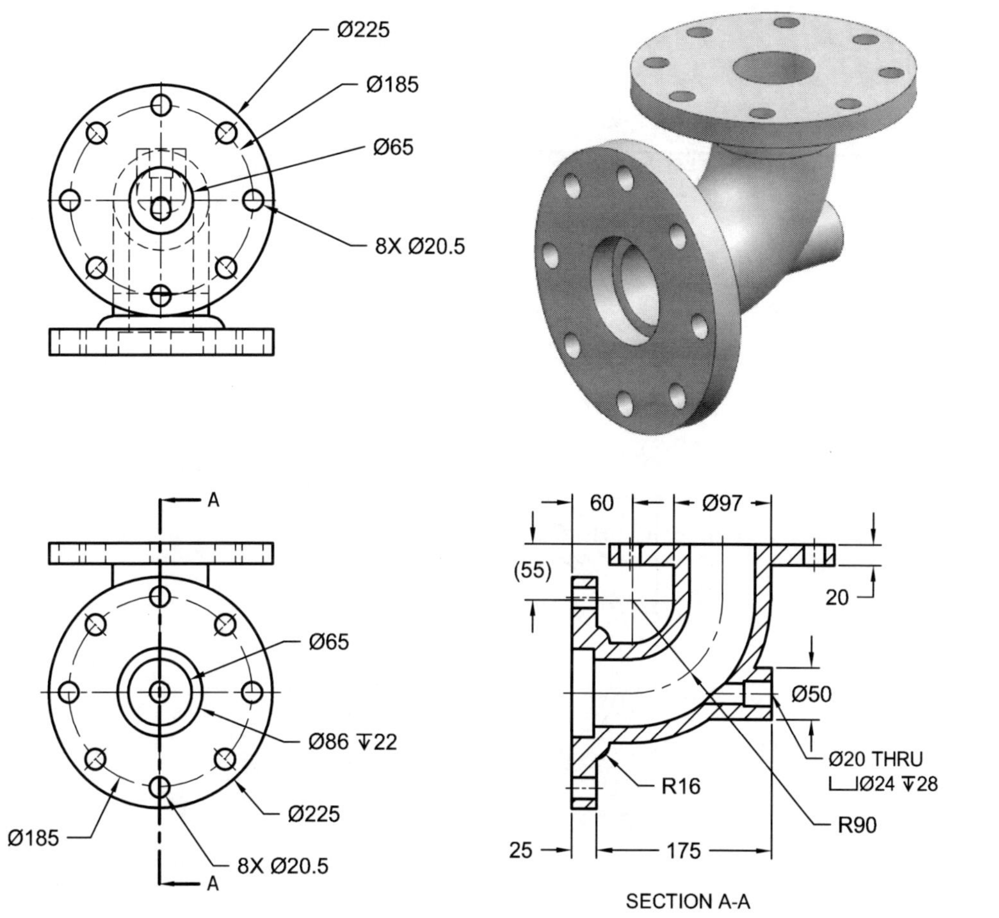

Figure 7-44 Model and dimensions for Exercise 1

Exercise 2

Create the model shown in Figure 7-45. The drawing views and dimensions of the model are shown in Figure 7-46. After creating the model, save it with the name *c07exr2.prt* at the location *\NX\c07.*

(Expected time: 30 min)

Figure 7-45 The solid model for Exercise 2

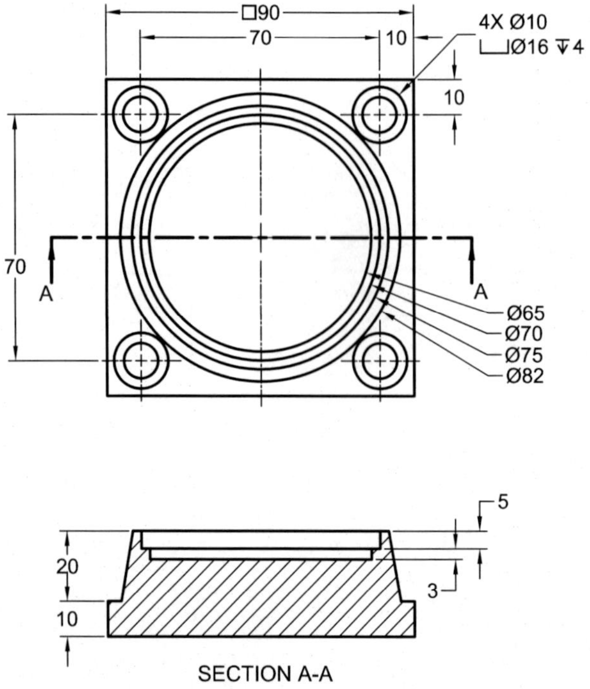

Figure 7-46 Views and dimensions for Exercise 2

Chapter 8

Assembly Modeling-I

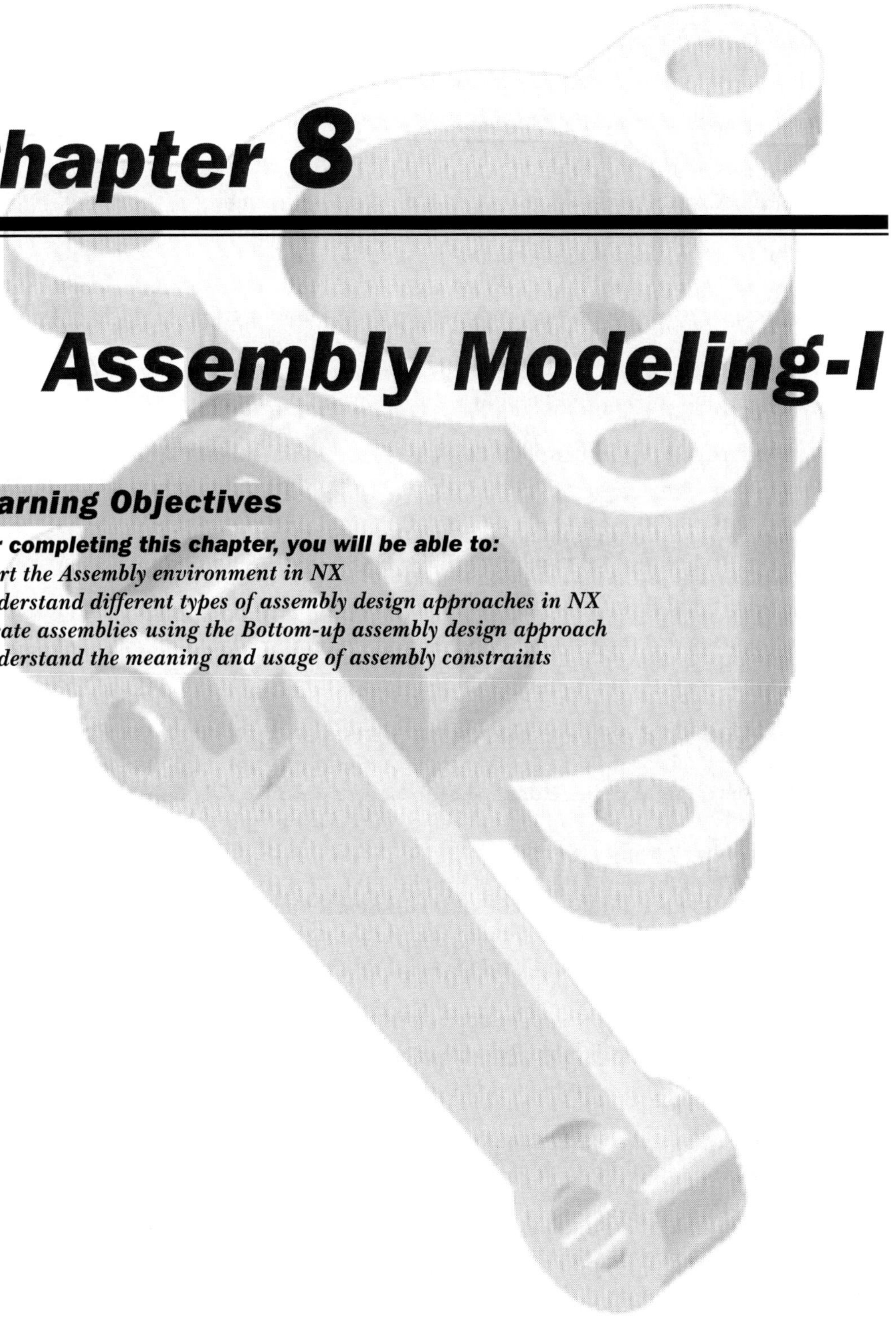

Learning Objectives

After completing this chapter, you will be able to:

- *Start the Assembly environment in NX*
- *Understand different types of assembly design approaches in NX*
- *Create assemblies using the Bottom-up assembly design approach*
- *Understand the meaning and usage of assembly constraints*

THE ASSEMBLY ENVIRONMENT

The Assembly environment is used to create interrelationship between the component parts which are assembled together by applying a parametric link, both in dimensional and positional aspects. The assembly constraints are parametric in nature so any types of modifications with the assembly constraints at any stage of the assembling procedure is possible. Figure 8-1 shows the Pipe Vice assembly created in the Assembly environment of NX. The assembly files in NX have the file name extension as .prt. The Assembly environment in NX is interactive and bidirectionally associative. The component can be modified in the assembly environment itself by setting the part to be modified as the work part. After assembling the components, you can also check the interference between them. This increases the efficiency of the assembly and also eliminates the errors while actually manufacturing the components. You can also create the exploded state of assembly. Creating this state of assembly helps the technician understand the sequence of components to be assembled better.

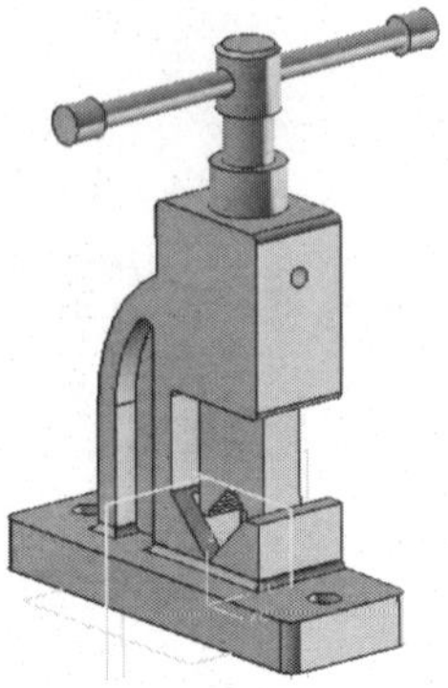

Figure 8-1 *The Pipe Vice assembly created in NX*

INVOKING THE ASSEMBLY ENVIRONMENT

You can invoke the Assembly environment by two methods:

1. Using the **Assembly** template from the **New** dialog box.
2. Invoking the Assembly environment in the current part file.

Invoking the Assembly Environment Using the New Dialog Box

To invoke the Assembly environment using the **Assembly** template, choose the **New** button from the **Standard** group of the **Home** tab; the **New** dialog box will be displayed. Choose the **Model** tab if it is not already chosen and select the **Assembly** template from the **Templates** rollout of the dialog box. Next, choose the button on the right side of the **Name** text box; the **Choose New File Name** dialog box will be displayed. Enter the assembly name in the **File name** edit box. Also, to specify the location to save the assembly file, browse to the folder where you need to save the assembly file. Choose the **OK** button twice; the Assembly environment with the **Add Component** dialog box will be invoked. The use of the **Add Component** dialog box is discussed later.

Invoking the Assembly Environment in the Current Part File

Ribbon: Application > Design > Assemblies

In the previous section, you learned to invoke the Assembly environment using the **New** dialog box. In this section, you will learn to invoke it in the current part file. To do so, choose the **Assemblies** tab from the **Ribbon**; the Assembly environment will be invoked. If the **Assemblies** tab is not already added to the **Ribbon**, then choose **Application > Design > Assemblies** from the **Ribbon**; the **Assemblies** tab will be added to the **Ribbon** and the **Assemblies** group will be added to the **Home** tab. The NX window with the Assembly environment is shown in Figure 8-2.

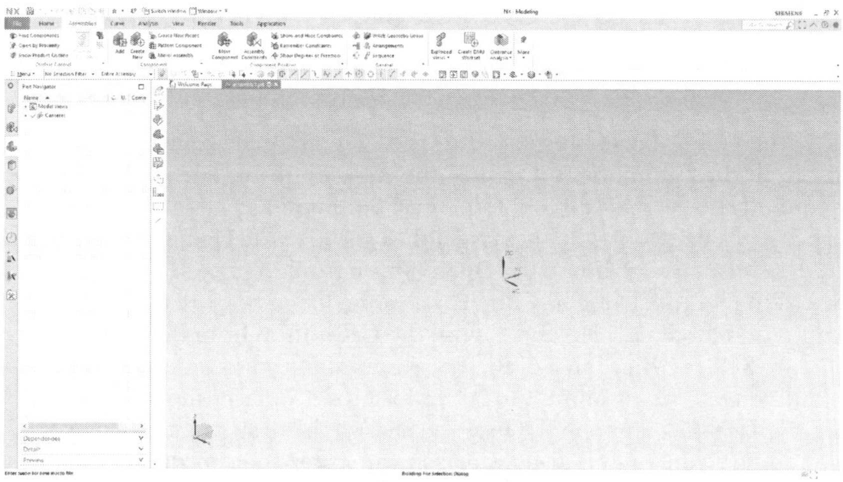

Figure 8-2 *The NX window in the Assembly environment*

Types of Assembly Design Approaches

In NX, the assemblies can be created using two types of design approaches, the bottom-up assembly design approach and the top-down assembly design approach. In this chapter, you will learn about the bottom-up assembly design approach. The top-down approach is discussed in the next chapter.

Bottom-up Assembly Design Approach

The Bottom-up assembly design approach is the most widely preferred traditional approach of assembly design. In this approach, all components are created as separate part files and placed in the assembly as external components. In this approach, you need to create components separately in the Modeling environment and save them as *.prt* files. After creating all components, you need to start a new file in the Assembly environment and place them in the assembly environment using the tools in the **Assemblies** group. Next, you need to assemble the components using assembly constraints.

The main advantage of using this approach is that you can pay more attention to the complexity of the component design as you create them before assembling. You can also capture the design intent easily. This approach is preferred while handling large assemblies or assemblies with complex components.

CREATING BOTTOM-UP ASSEMBLIES

As mentioned earlier, in the bottom-up assemblies, components are created as separate parts in the Modeling environment, and then they are placed in a new file that is started in the Assembly environment. In the assembly file, the parts are assembled using the assembly constraints. As discussed earlier, it is recommended that you first invoke the Modeling environment and then choose **Application** > **Design** > **Assemblies** from the **Ribbon** to invoke the Assembly environment.

Placing Components in the Assembly Environment

Ribbon: Assemblies > Component > Add
Menu: Assemblies > Components > Add Component

Add

In NX, the components are placed in the Assembly environment using the **Add** tool available in the **Component** group of the **Assemblies** tab in the **Ribbon**. To place the component, choose the **Add** tool from the **Component** group of **Assemblies** tab. You can also choose the **Add** tool from the **Assemblies** group of the **Home** tab; the **Add Component** dialog box will be displayed. Choose the **Open** button from the **Loaded Parts** sub-rollout of the **Part to Place** rollout of the dialog box; the **Part Name** dialog box will be displayed. Browse and select the component to be placed and choose the **OK** button from the **Part Name** dialog box. Next, click in the **Select Object** area of the **Location** rollout; the component will be attached to the cursor and you will be prompted to select location to place the component. Click in the drawing area to place the component. Once you click in the drawing area, a dynamic triad along with the input edit boxes will be displayed. You can enter the coordinate values in these edit boxes to place the component. The Assembly environment window with the **Add Component** dialog box is shown in Figure 8-3.

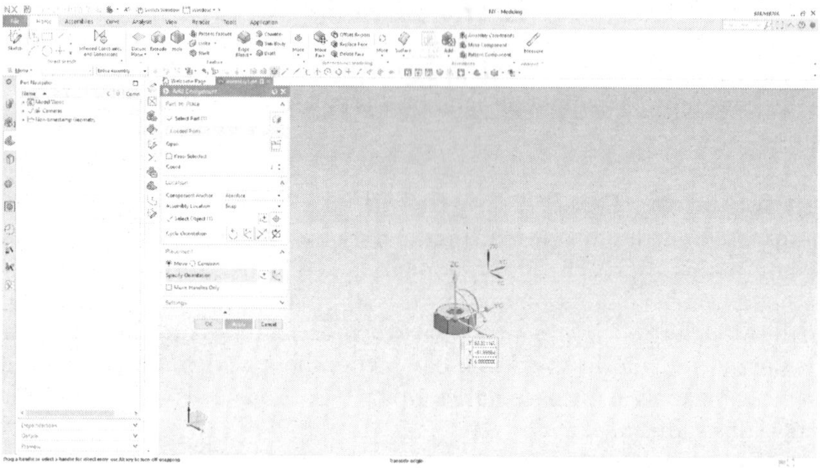

***Figure 8-3** The Assembly environment window with the **Add Component** dialog box*

Note

*While placing the component in the drawing area, a dynamic triad along with the input edit boxes are displayed on the component. This is because the **Move** radio button is selected in the **Placement** rollout of the **Add Component** dialog box. You can also select the **Constrain** radio button in the **Placement** rollout. On doing so, the **Constraint Type** sub-rollout will be displayed. You can use the constraint buttons available in this sub-rollout to place the components.*

The name of the component is displayed in the **File Name** list of the **Loaded Parts** sub-rollout of the **Part to Place** rollout of the **Add Component** dialog box.

Note

If the component to be assembled is already placed in the assembly, then for subsequent placements, you can directly retrieve the component from the ***File Name*** *list of the* ***Loaded Parts*** *sub-rollout of the* ***Part to Place*** *rollout. To do so, invoke the* ***Add Component*** *dialog box and then select the component to be placed from the* ***Loaded Parts*** *sub-rollout. Next, click in the* ***Select Object*** *area of the* ***Location*** *rollout and follow the same procedure explained before to place the component.*

You can add multiple instances of the same component to the assembly file. To do so, enter the number of instances in the **Count** edit box available in the **Part to Place** rollout. The remaining options in the **Add Component** dialog box are discussed next.

Location Rollout

The options available in this rollout are discussed next.

Component Anchor

This drop-down list lists all the possible anchors for the location of a component. By default, the **Absolute** option is selected in this drop-down list which takes the absolute origin of the component as the anchor.

Assembly Location

The options in this drop-down list enable you to locate the components in the drawing window. The **Snap** option allows you to select the placement face based on the orientation of the assembly and the location of the cursor. If you select the **Absolute - Work Part** option from this drop-down list, the component anchor will be placed at the absolute origin of the work part. If you select the **Absolute - Displayed Part** option, the component anchor will be placed at the absolute origin of the displayed part. The **WCS** option enables you to place the component anchor at the current WCS location and orientation.

Cycle Orientation

This area of the **Location** rollout contains four buttons which enable you to specify different component orientations based on the location settings of the component.

Placement Rollout

This rollout contains two radio buttons. The function of each radio button has been discussed earlier.

Settings Rollout

The options in this rollout are discussed next.

Enabled Constraints

This sub-rollout will be available only if the **Constrain** radio button is selected in the **Placement** rollout. Using this sub-rollout, you can control the availability of constraints in the **Constraint Type** sub-rollout of the **Placement** rollout. If you select the check box available on the left of any constraint in this sub-rollout, the corresponding constraint will be available in the **Constraint Type** sub-rollout of the **Placement** rollout.

Interaction Options

This sub-rollouts contains four check boxes. On selecting the **Scatter Components** check box, the components are automatically placed at different locations so that they do not overlap. The **Keep Constraints** check box is selected to create the constraints used to place the component. You can select these constraints from the **Assembly Navigator** and the **Constraint Navigator**. On selecting the **Preview** check box, the component is displayed in the drawing area while placing it. On selecting the **Preview Window** check box, a preview of the component is displayed in a separate **Component Preview** window.

Component Name

This edit box is used to display the name of the currently selected component in the **Add Component** dialog box.

Reference Set

The options in the **Reference Set** drop-down list are used to specify the state of the component that will occur in the assembly. If you select the **Model** option from the **Reference Set** drop-down list, only the model will be placed in the assembly file. If you select the **Entire Part** option, the model will be placed in the assembly file along with the datum planes and the sketches used for creating it. If you select the **Empty** option, the component will be placed in the assembly file as an empty part without the model or its reference sets.

Layer Option

The options in this drop-down list are used to specify different layers in the assembly file.

Tip

*It is recommended to assemble the base component by selecting the **Entire Part** option from the **Reference Set** drop-down list.*

TUTORIAL

Tutorial 1

In this tutorial, you will create the components of the Plummer Block assembly shown in Figure 8-4 and then assemble them. The dimensions of the components are given in Figures 8-5 through 8-7. Save the file with the name *Plummer Block.prt* at the location *NX\c08\Plummer Block*.

(Expected time: 2.5 hrs)

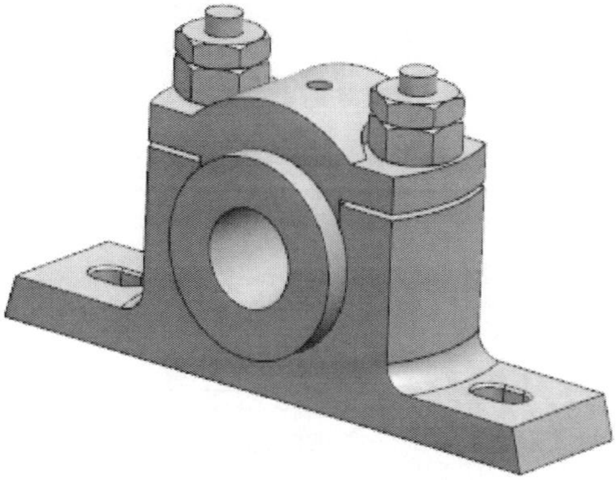

Figure 8-4 *The Plummer Block assembly*

The following steps are required to complete this tutorial:

a. Create all components as individual part files and save them in the *Plummer Block* folder at the location *\NX\c08*.
b. Start a new file and invoke the Assembly environment.
c. Insert the Casting in the assembly window and assemble it, refer to Figure 8-8.
d. Insert the Brasses in the assembly window and apply the required assembly constraints to make it a fully-constrained part, refer to Figures 8-9 through 8-13.
e. Insert the Cap in the assembly window and apply the required assembly constraints, refer to Figures 8-14 through 8-18.
f. Assemble the Bolts by using the required assembly constraints, refer to Figures 8-19 through 8-23.
g. Assemble the Nuts by using the required assembly constraints, refer to Figures 8-24 through 8-28.
h. Assemble the Lock Nuts by using the required assembly constraints, refer to Figures 8-29 through 8-32.

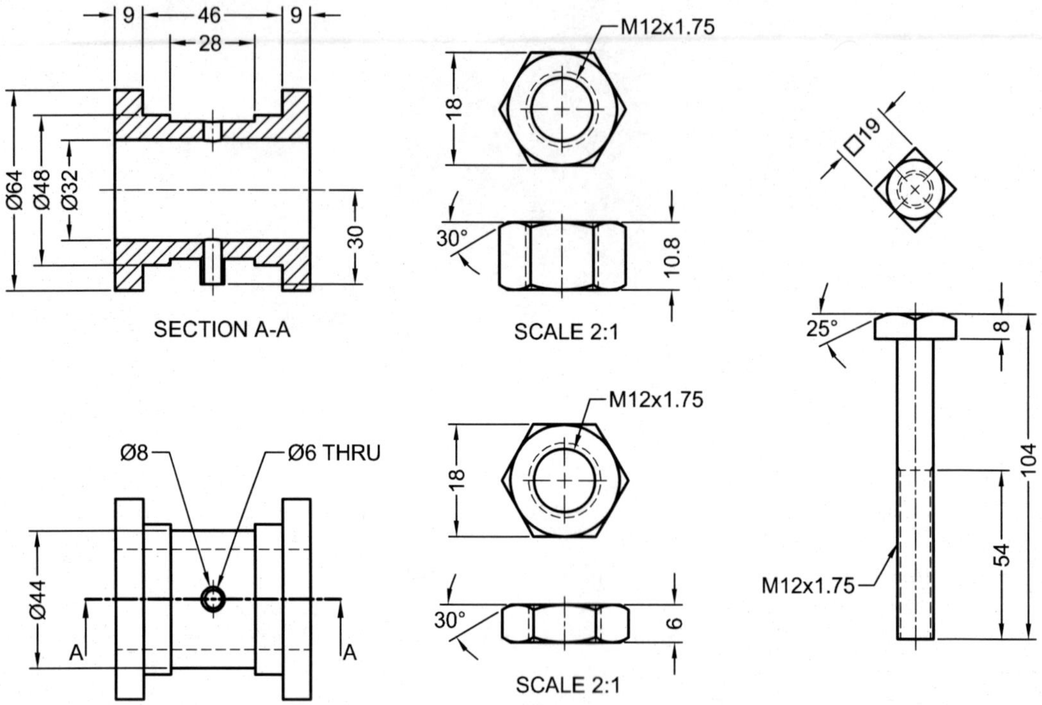

Figure 8-5 *Views and dimensions of Brasses, Nut, Lock Nut, and Bolt*

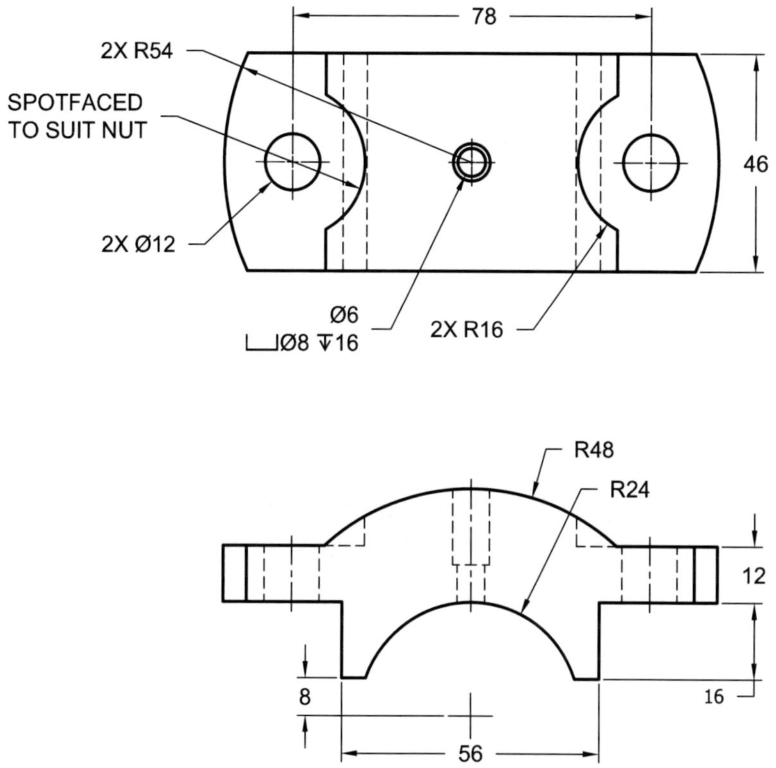

Figure 8-6 *Views and dimensions of the Cap*

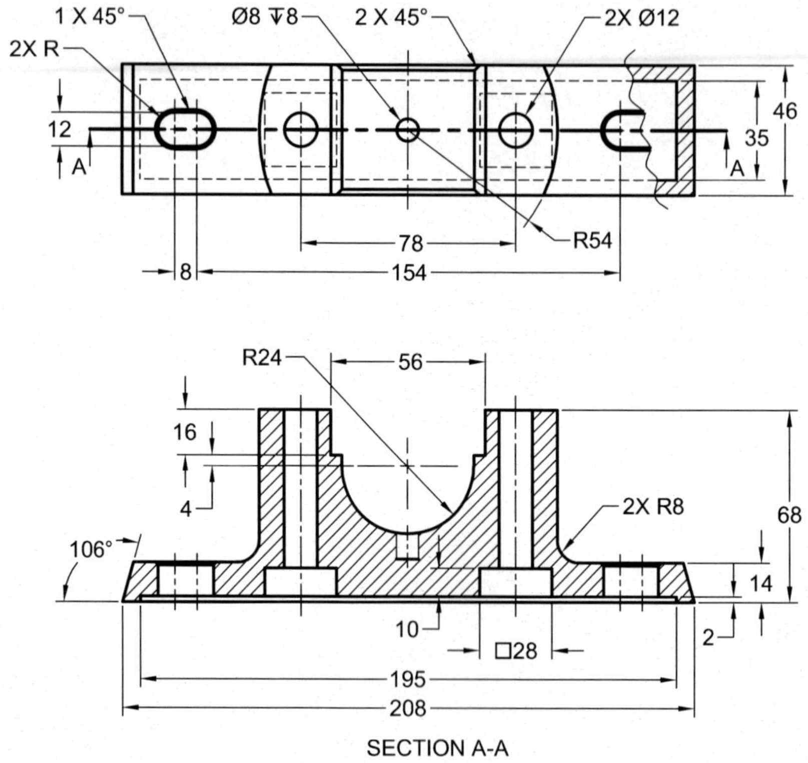

Figure 8-7 Views and dimensions of the Casting

Creating Assembly Components

1. Create all components of the Plummer Block assembly as separate part files. Specify the names of the files as mentioned in the drawing views. Save the files at the location *\NX\c08\Plummer Block*.

Starting a New Assembly File

1. Start a new file with the name *Plummer Block.prt* by using the **Model** template and specify its location as *C:\NX\c08\Plummer Block.*

2. Invoke the Assembly environment, if it is not already invoked, by choosing the **Assemblies** tab from the **Ribbon**.

Assembling the Casting with the Datum Planes

1. Choose the **Add** tool from the **Assemblies** tab; the **Add Component** dialog box is displayed.

2. Choose the **Open** button from the **Add Component** dialog box; the **Part Name** dialog box is displayed.

3. Double-click on the Casting in the **Part Name** dialog box. Next, select the **Entire Part** option from the **Reference Set** drop-down list in the **Settings** rollout.

4. Click in the **Select Object** area of the **Location** rollout; the Casting is attached to the cursor and you are prompted to select location to place the component. Click anywhere in the drawing area to place the component. Select the **Move** radio button from the **Placement** rollout, if not already selected; the **X**, **Y**, and **Z** edit boxes are displayed in the drawing area. Enter **0**, **0**, and **0** in the respective edit boxes and then press ENTER; the absolute origin of the component is placed at the origin of the Datum CSYS.

5. Choose the **Apply** button from the **Add Component** dialog box; the **Create Fix Constraint** window is displayed. Choose the **Yes** button from it; the **Fix** constraint is applied to the Casting and all its degrees of freedom are restricted.

6. Right-click on the Casting to display a shortcut menu. Next, choose **Replace Reference Set > MODEL** from the shortcut menu. The assembly after assembling the Casting is shown in Figure 8-8.

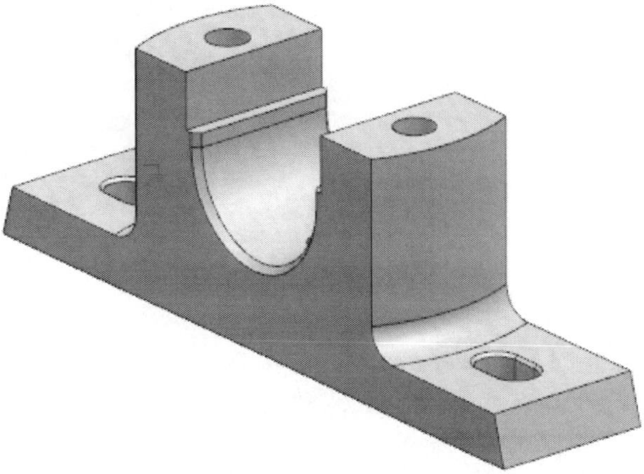

Figure 8-8 *Casting assembled in the Assembly environment*

Note

*In this tutorial, the display of the constraints has been turned off. To do so, click on the **Assembly Navigator** tab in the **Resource Bar**; the **Assembly Navigator** cascading menu is displayed. Right-click on the **Constraints** node; a shortcut menu is displayed. Next, clear the check mark on the left of the **Display Constraints in Graphics Window** option by choosing it.*

Assembling the Second Component

After inserting the Casting, you need to assemble Brasses as the second component.

1. Choose the **Add** tool from the **Assemblies** tab; the **Add Component** dialog box is displayed.

2. Choose the **Open** button from the **Add Component** dialog box; the **Part Name** dialog box is displayed.

3. Double-click on the Brasses in the **Part Name** dialog box. Next, select the **Model ("MODEL")** option from the **Reference Set** drop-down list in the **Settings** rollout.

4. Next, click in the **Select Object** area of the **Location** rollout; the Brasses is attached to the cursor and you are prompted to select location to place the component. Click anywhere in the drawing area to place the component. Select the **Constrain** radio button in the **Placement** rollout; the **Constrain Type** sub-rollout is displayed.

5. Choose the **Touch Align** option from the **Constraint Type** sub-rollout and then select the **Infer Center/Axis** option from the **Orientation** drop-down list in the **Geometry to Constrain** sub-rollout.

6. Select the cylindrical face of the Brasses, refer to Figure 8-9, and then select the cylindrical face of the Casting, refer to Figure 8-10.

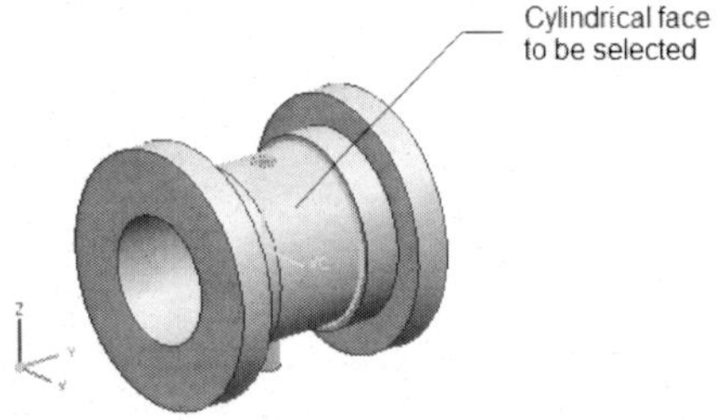

Figure 8-9 *Cylindrical face to be selected from the Brasses*

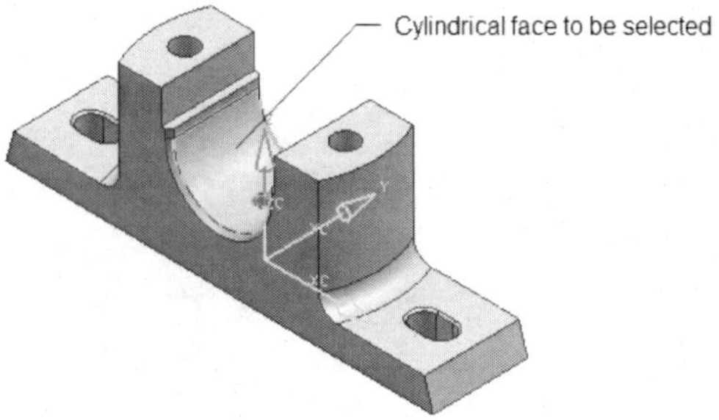

Figure 8-10 *Cylindrical face to be selected from the Casting*

7. Next, select the **Fit** option from the **Constraint Type** sub-rollout of the **Add Component** dialog box.

8. Select the cylindrical face of the Brasses, refer to Figure 8-11, and then select the cylindrical face of the Casting, refer to Figure 8-12.

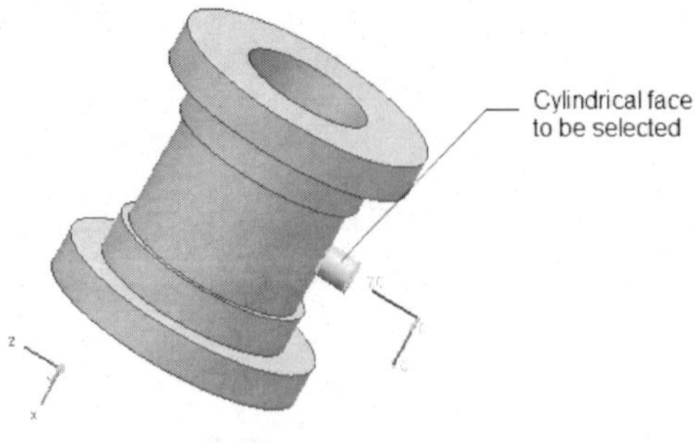

Figure 8-11 *Cylindrical face to be selected from the Brasses*

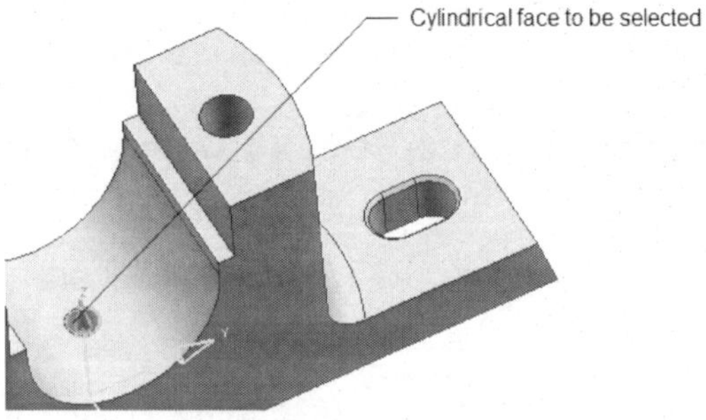

Figure 8-12 *Cylindrical face to be selected from the Casting*

9. Next, choose the **Apply** button from the **Add Component** dialog box.

   The assembly after assembling the Brasses is shown in Figure 8-13.

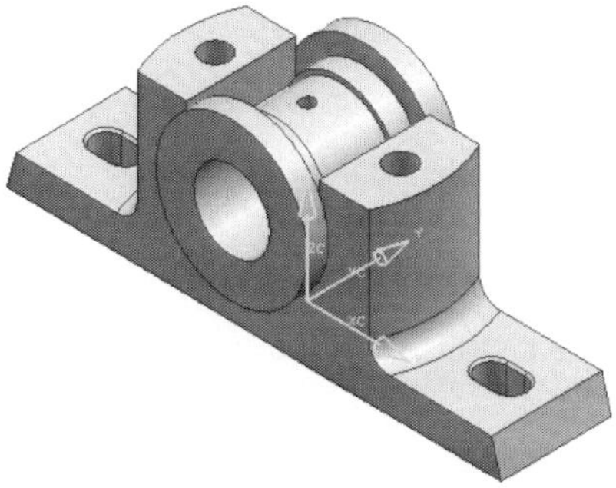

Figure 8-13 *Assembly after assembling the Brasses*

Assembling the Third Component

Next, you need to assemble the Cap.

1. Choose the **Open** button from the **Add Component** dialog box; the **Part Name** dialog box is displayed.

2. Double-click on the Cap in the **Part Name** dialog box.

3. Next, click in the **Select Object** area of the **Location** rollout; the Cap is attached to the cursor and you are prompted to select location to place the component. Click anywhere in the drawing area to place the component.

4. Choose the **Touch Align** option from the **Constraint Type** sub-rollout and then select the **Infer Center/Axis** option from the **Orientation** drop-down list in the **Geometry to Constrain** sub-rollout.

5. Select the cylindrical face of the Cap, refer to Figure 8-14. Next, select the cylindrical face of the Brasses, refer to Figure 8-15.

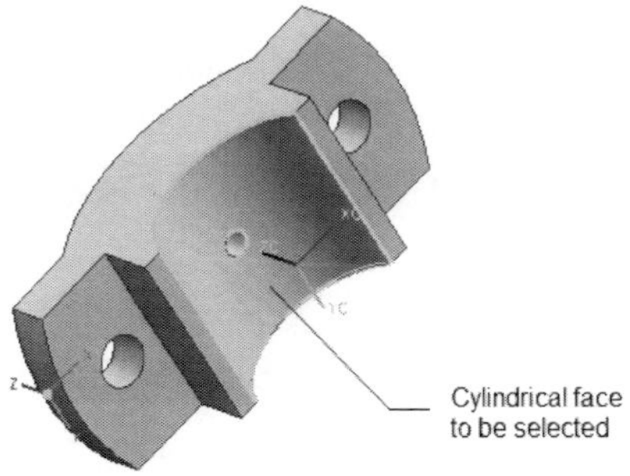

Figure 8-14 *Cylindrical face to be selected from the Cap*

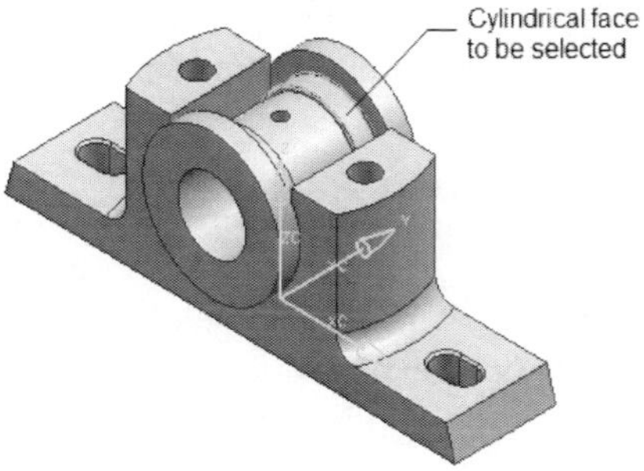

Figure 8-15 *Cylindrical face to be selected from the Brasses*

6. Select the cylindrical face of the Cap, refer to Figure 8-16. Next, select the cylindrical face of the Brasses, refer to Figure 8-17.

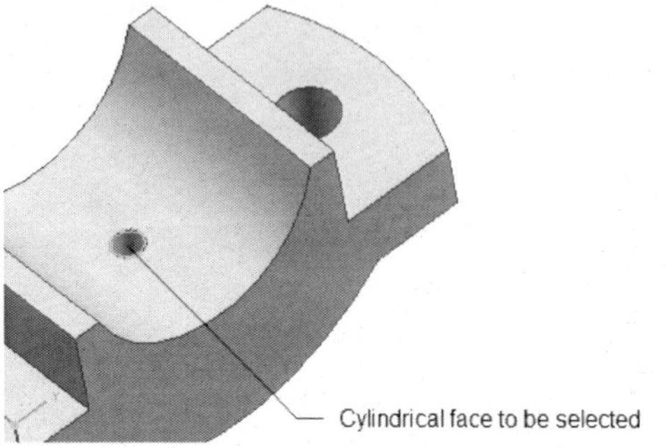

Figure 8-16 *Cylindrical face to be selected from the Cap*

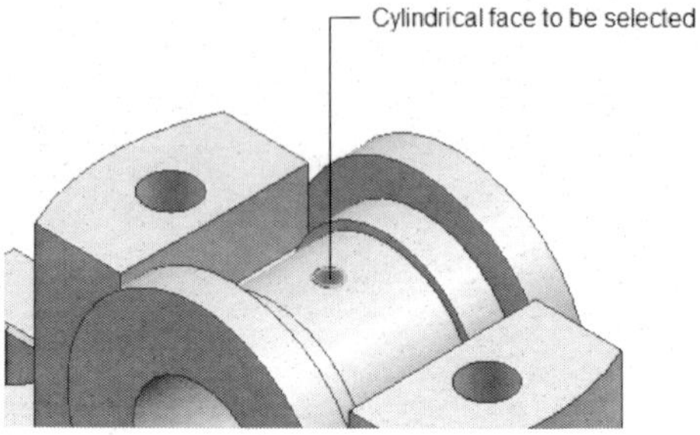

Figure 8-17 *Cylindrical face to be selected from the Brasses*

7. Next, choose the **Apply** button from the **Add Component** dialog box. The resulting assembly after assembling the Cap is shown in Figure 8-18.

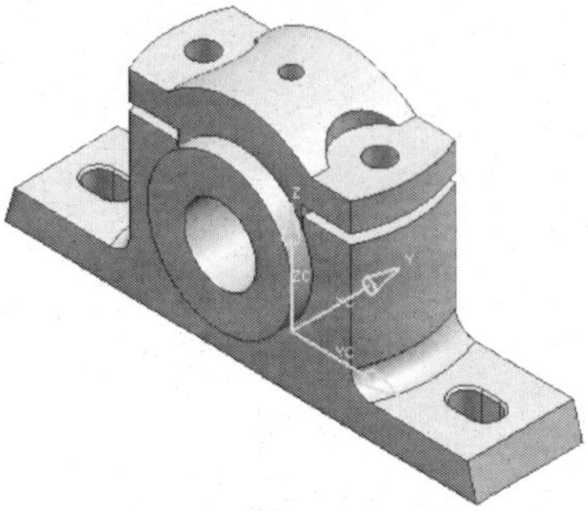

Figure 8-18 *The assembly after assembling the Cap*

Assembling the Bolts

Next, you need to assemble the Bolt as the fourth component.

1. Choose the **Open** button from the **Add Component** dialog box; the **Part Name** dialog box is displayed.

2. Double-click on the Bolt in the **Part Name** dialog box.

3. Next, click in the **Select Object** area of the **Location** rollout; the Bolt is attached to the cursor and you are prompted to select location to place the component. Click anywhere in the drawing area to place the component.

4. Choose the **Align/Lock** option from the **Constraint Type** sub-rollout.

5. Select the cylindrical face of the Bolt, refer to Figure 8-19. Next, select the cylindrical face from the Casting, refer to Figure 8-20.

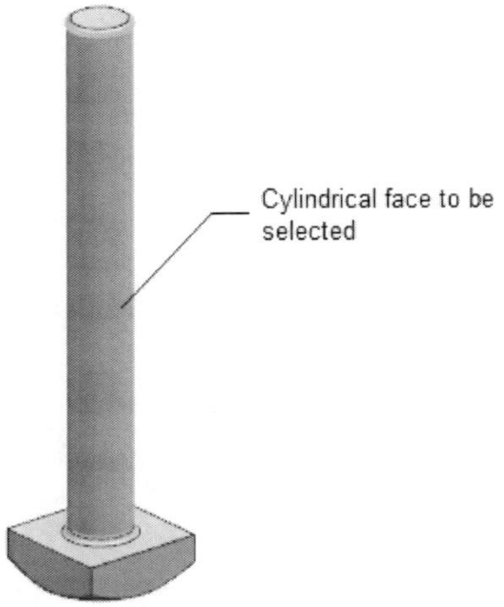

Figure 8-19 *Cylindrical face to be selected from the Bolt*

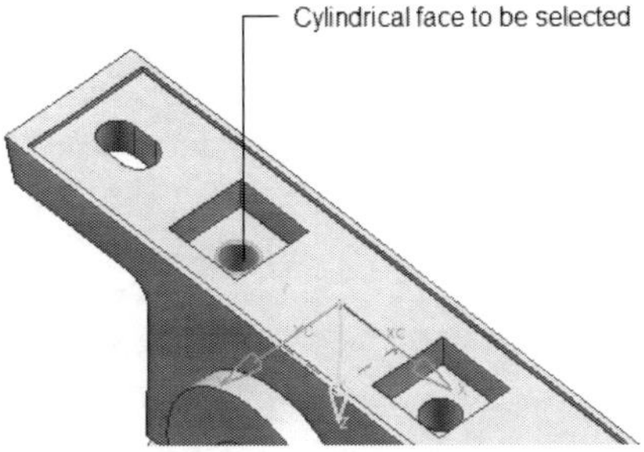

Figure 8-20 *Cylindrical face to be selected from the Casting*

6. Choose the **Touch Align** option from the **Constraint Type** sub-rollout and the **Touch** option from the **Orientation** drop-down list in the **Geometry to Constrain** sub-rollout.

7. Select the planar face of the Bolt, refer to Figure 8-21. Next, select the planar face of the Casting, refer to Figure 8-22.

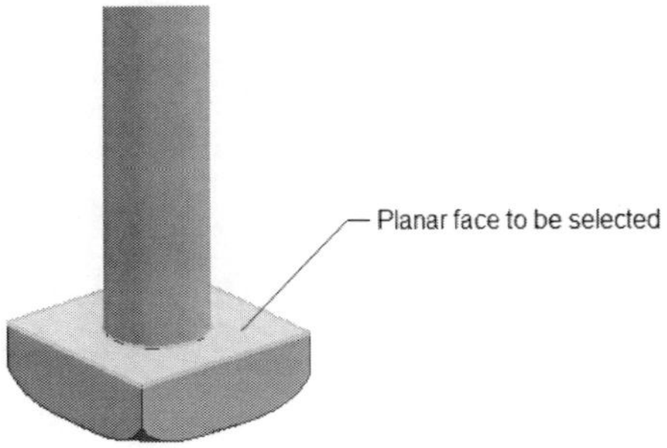

Figure 8-21 *Planar face to be selected from the Bolt*

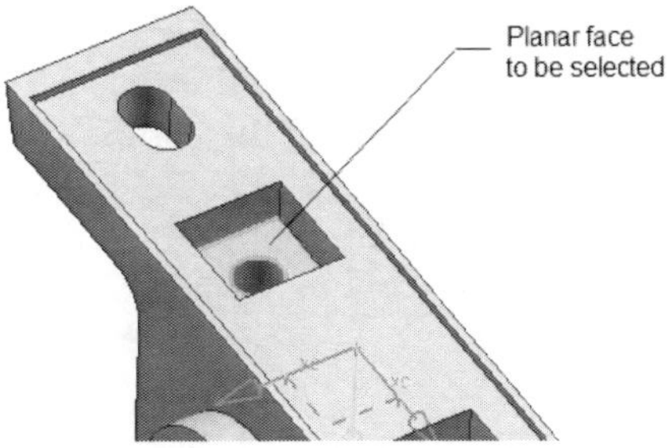

Figure 8-22 *Planar face to be selected from the Casting*

8. Next, choose the **Apply** button from the **Add Component** dialog box.

9. Similarly, assemble the second instance of the Bolt on the other side. The assembly after assembling the bolts is shown in Figure 8-23.

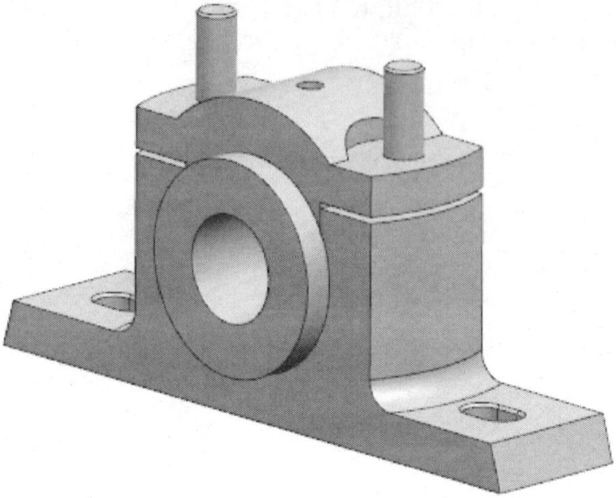

Figure 8-23 *The assembly after assembling the Bolts*

Note
*You might need to use the **Parallel** constraint to align the bolt head in the casting rectangular slot.*

Assembling the Nuts

Next, you need to assemble the Nut as the fifth component.

1. Choose the **Open** button from the **Add Component** dialog box; the **Part Name** dialog box is displayed.

2. Double-click on the Nut in the **Part Name** dialog box.

3. Next, click in the **Select Object** area of the **Location** rollout; the Nut is attached to the cursor and you are prompted to select location to place the component. Click anywhere in the drawing area to place the component.

4. Choose the **Touch Align** option from the **Constraint Type** sub-rollout and then select the **Infer Center/Axis** option from the **Orientation** drop-down list in the **Geometry to Constrain** sub-rollout.

5. Select the cylindrical face of the Nut, refer to Figure 8-24. Next, select the cylindrical face of the Bolt, refer to Figure 8-25.

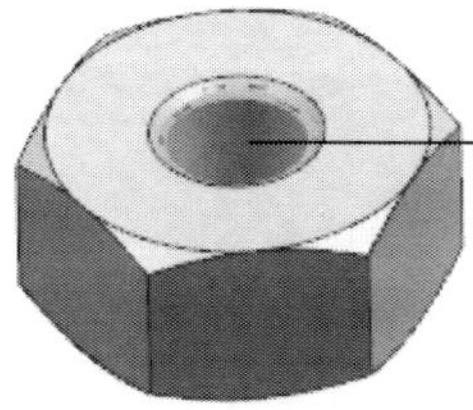

Figure 8-24 *Cylindrical face to be selected from the Nut*

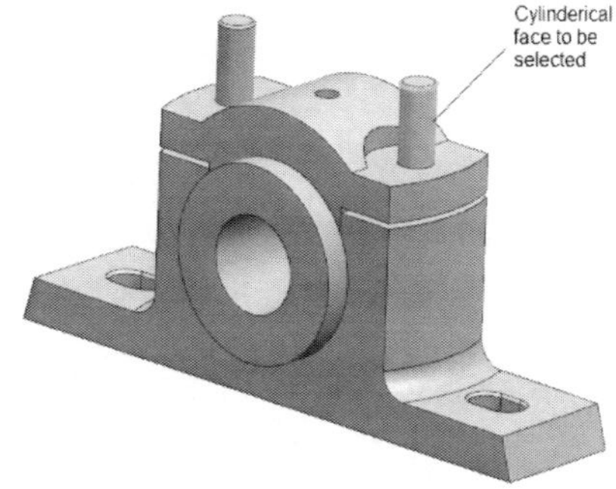

Figure 8-25 *Cylindrical face to be selected from Bolt*

6. Next, select the **Touch** option from the **Orientation** drop-down list.

7. Rotate the view of the Nut and select its bottom planar face, refer to Figure 8-26. Next, select the planar face from the Cap, refer to Figure 8-27.

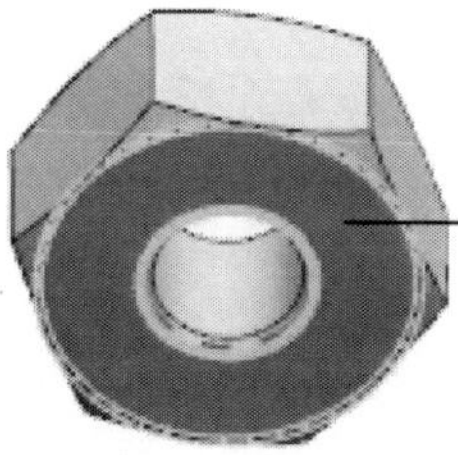

Figure 8-26 *Bottom planar face to be selected from the Nut*

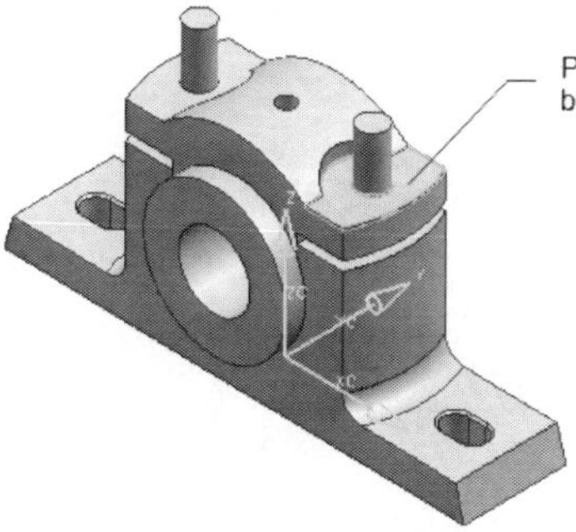

Figure 8-27 *Planar face to be selected from the Cap*

Note

You might need to turn off the display of the Casting while assembling the nut.

8. Choose the **Apply** button from the **Add Component** dialog box.

9. Similarly, assemble the second instance of the Nut on the other side. The assembly after assembling the Nuts is shown in Figure 8-28.

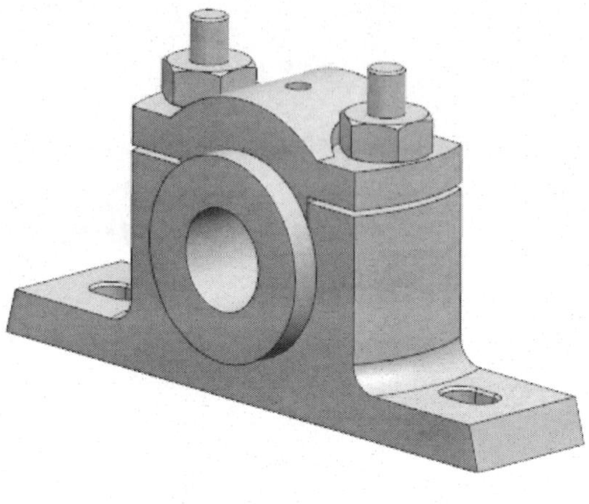

Figure 8-28 *The assembly after assembling the Nuts*

Tip

While assembling the Nut in the assembly, you should not constrain its rotational degree of freedom because on doing so, it will not be able to pass through the bolt threads to lock the Cap and Brasses with the Casting.

Assembling the Lock Nut

Next, you need to assemble the Lock Nut as the sixth component.

1. Choose the **Open** button from the **Add Component** dialog box; the **Part Name** dialog box is displayed.

2. Double-click on the Lock Nut in the **Part Name** dialog box.

3. Next, click in the **Select Object** area of the **Location** rollout; the Lock Nut is attached to the cursor and you are prompted to select location to place the component. Click anywhere in the drawing area to place the component.

4. Select the **Touch Align** option from the **Constraint Type** sub-rollout and then the **Touch** option from the **Orientation** drop-down list.

5. Select the planar face of the Lock Nut, refer to Figure 8-29. Next, select the planar face of the Nut, refer to Figure 8-30.

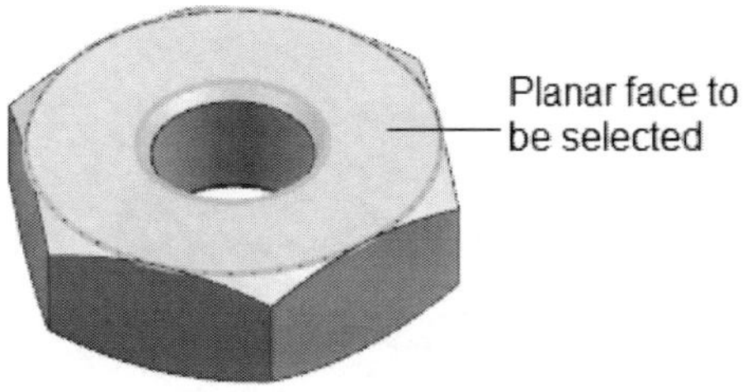

Figure 8-29 *Planar face to be selected from the Lock Nut*

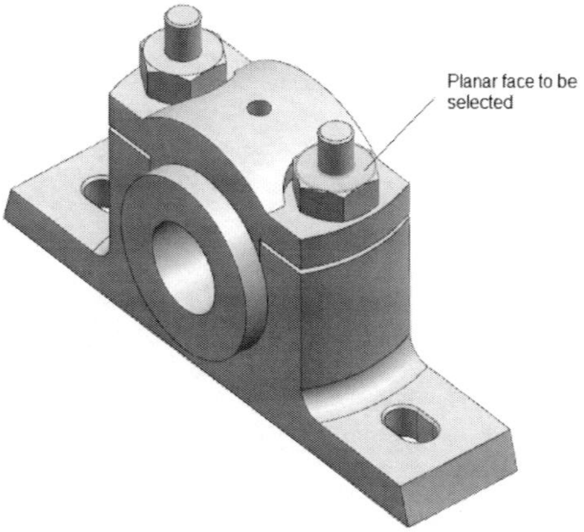

Figure 8-30 *Planar face to be selected from the Nut*

6. Select the **Infer Center/Axis** option from the **Orientation** drop-down list.

7. Select the cylindrical face of the Lock Nut, as shown in Figure 8-31. Next, select the cylindrical face of the Bolt, as shown in Figure 8-32.

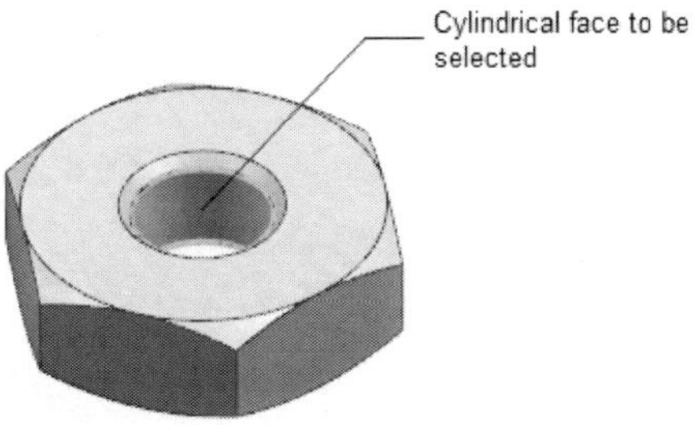

Figure 8-31 *Cylindrical face to be selected from the Lock Nut*

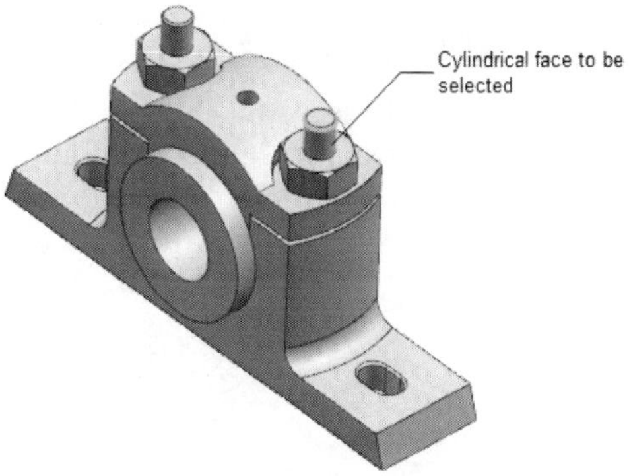

Figure 8-32 *Cylindrical face to be selected from the Bolt*

8. Choose the **Apply** button from the **Add Component** dialog box.

9. Similarly, assemble the second instance of the Lock Nut on the other side and then choose the **OK** button to exit the **Add Component** dialog box. The completed Plummer Block assembly is shown in Figure 8-33.

Figure 8-33 *The final Plummer Block assembly*

Saving and Closing the File

1. Choose **Menu > File > Close > Save and Close** from the **Top Border Bar** to save and close the file.

EXERCISE

Exercise 1

Create the Double Bearing assembly shown in Figure 8-34. The dimensions of the components of the assembly are given in Figures 8-35 through 8-38. **(Expected time: 4 hrs)**

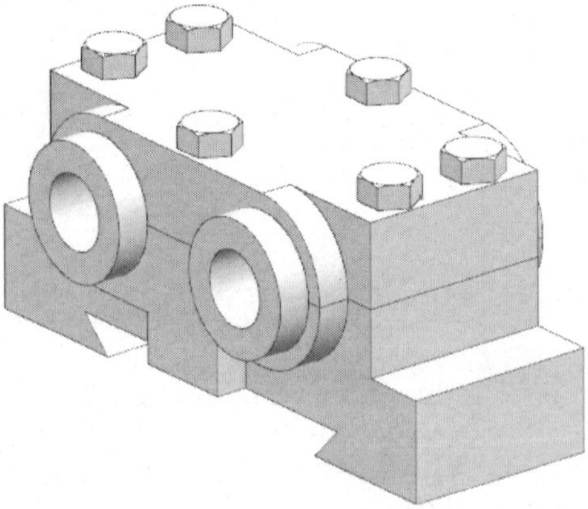

Figure 8-34 *The Double Bearing assembly*

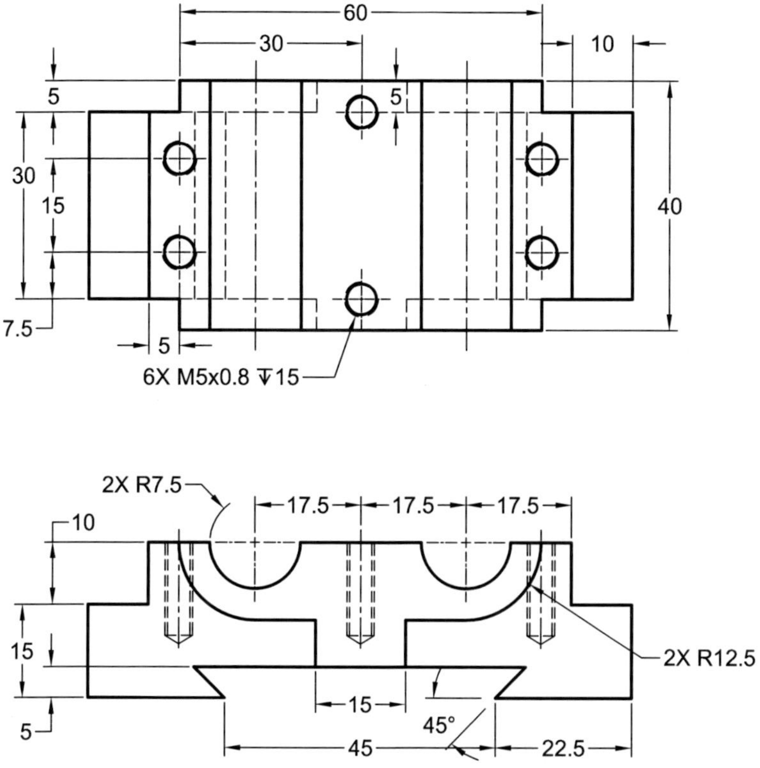

Figure 8-35 Orthographic views of the Base

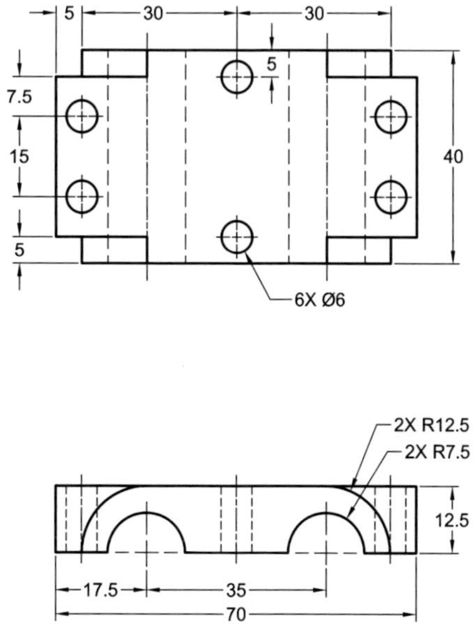

Figure 8-36 Orthographic views of the Cap

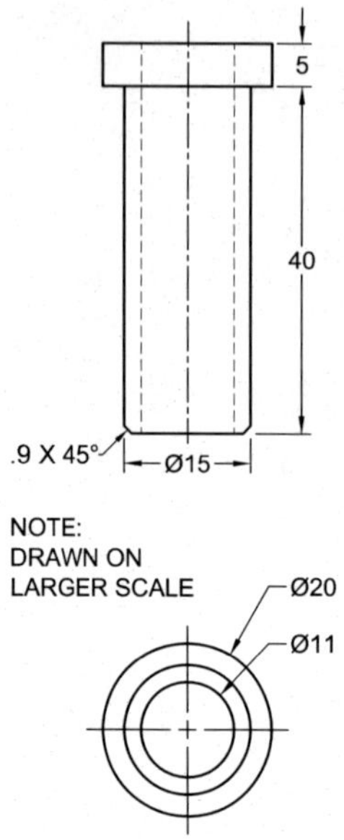

***Figure 8-37** Orthographic views of the Bushing*

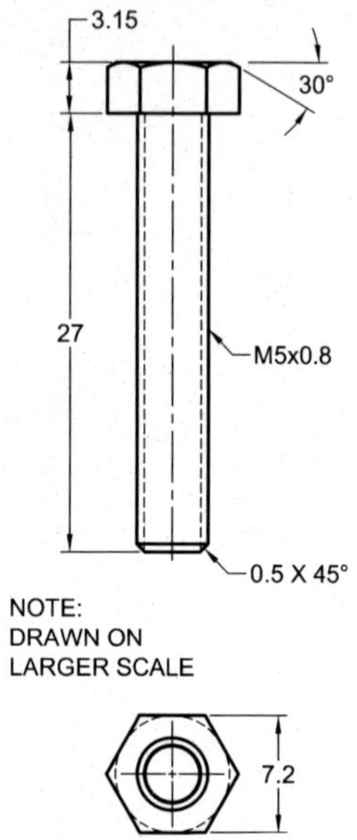

***Figure 8-38** Orthographic views of the Bolt*

Chapter 9

Assembly Modeling-II

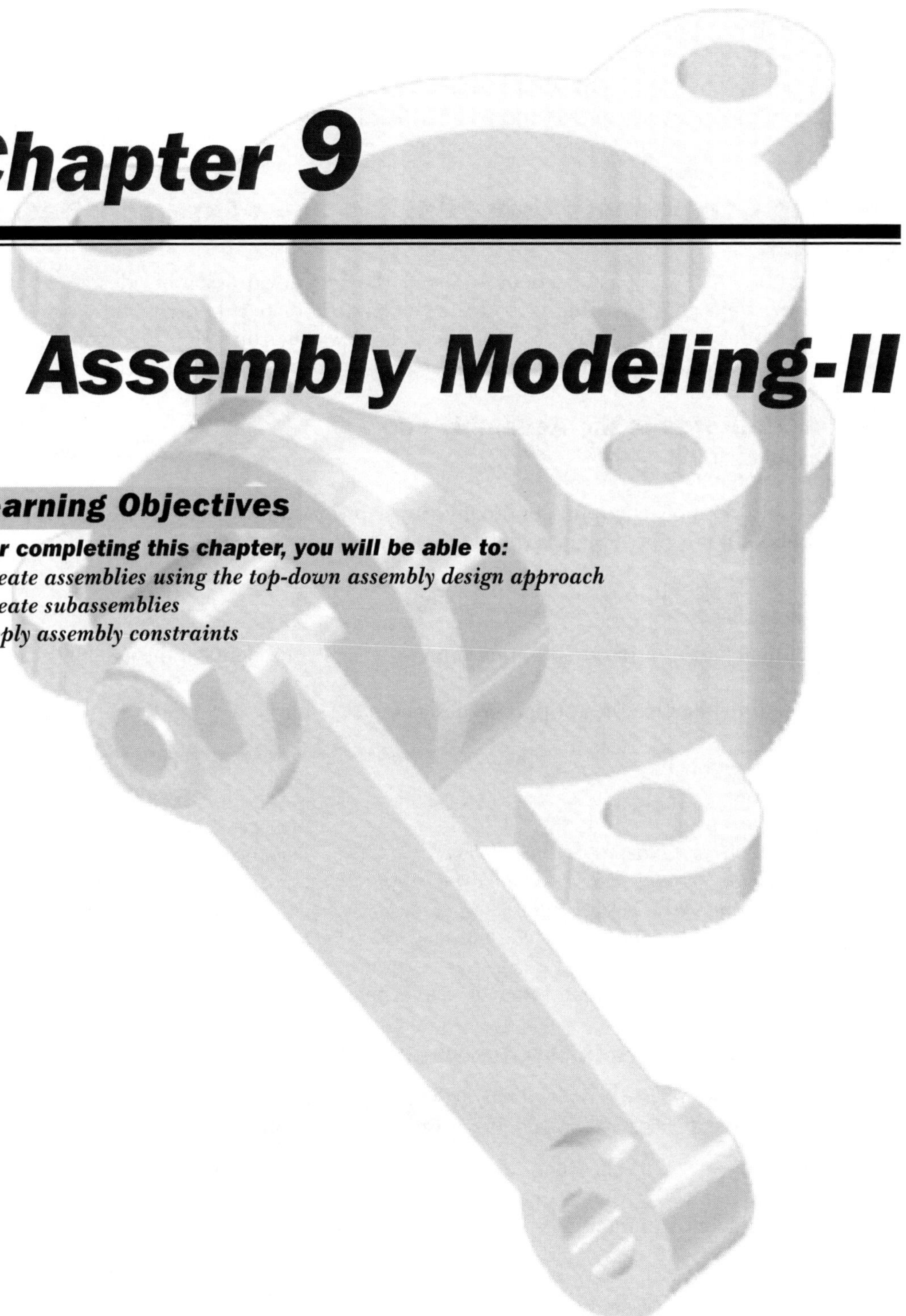

Learning Objectives

After completing this chapter, you will be able to:

- *Create assemblies using the top-down assembly design approach*
- *Create subassemblies*
- *Apply assembly constraints*

THE TOP-DOWN ASSEMBLY DESIGN APPROACH

As discussed in the previous chapter, you can also create all the components of an assembly in a single file as separate bodies or features and then save them as separate parts. This approach of creating an assembly is known as the top-down assembly design approach. In this approach, most of the dimensions are derived from other features and part bodies. The procedure for creating an assembly using the top-down assembly design approach is discussed next.

Creating Components Using the Top-down Assembly Design Approach

The procedure for creating an assembly using the top-down assembly design approach involves three steps. The first step is to create the features or part bodies. The second step is to save the features or part bodies as separate part files. The third step is to apply assembly constraints to the components and create the assembly. These steps are discussed next.

Creating Features in the Assembly File

1. Start a new part file.

2. Invoke the Assembly environment by choosing **Application > Design > Assemblies** from the **Ribbon**, if it is not already invoked.

3. Choose the **Assemblies** tab from the **Ribbon.**

4. Create the features or part bodies that represent the components of the assembly.

5. Save the individual features or part bodies as separate parts.

6. Save the assembly file.

Note

You can create features as separate bodies or in union with an existing feature. For reorienting the WCS of the sketching plane, you should only use the datum axis of the assembly file as the reference entity. If you select reference entities such as an edge or a datum axis of the existing features, then the feature containing the reference entity will also be added to the feature created.

The procedure for saving the features created as separate part files is discussed next.

Creating Components in the Assembly File

Ribbon:	Assemblies > Component > Create New
Menu:	Assemblies > Components > Create New Component

Create New

You can create different components in the assembly environment and then save them as separate part files one by one. The procedure for creating and saving the components in assembly environment as separate part files is discussed next.

1. Choose the **Create New** tool from the **Component** group in the **Assemblies** tab; the **New Component File** dialog box will be displayed and you will be prompted to select a template. Select the **Model** template from the **Templates** rollout.

2. Specify the name and location of the file in the **New File Name** rollout of the **New Component File** dialog box.

3. Choose the **OK** button from the dialog box; the **Create New Component** dialog box will be displayed, as shown in Figure 9-1.

   By default, the file name that is specified in step 2 in the **New Component File** dialog box will be displayed in the **Component Name** text box of the **Settings** rollout.

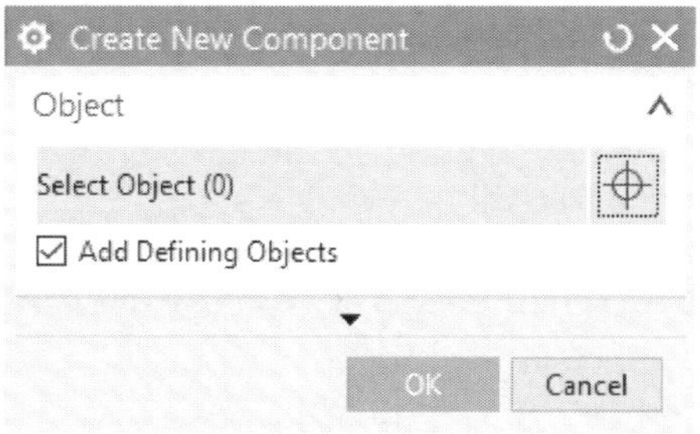

Figure 9-1 *The* ***Create New Component*** *dialog box*

4. By default, the **Select Objects** button is chosen in the **Create New Component** dialog box and you will be prompted to select objects for the new component. If you have created any object, you can select it to include in the new part file. Else, you can create an empty part file.

   Also, the **Add Defining Objects** check box is selected by default in the **Object** rollout of the **Create New Component** dialog box. As a result, the defined objects such as datum planes, axes, or points of the selected object will be included in the new part file. To exclude the defined objects from the new part file, clear the **Add Defining Objects** check box.

5. Now, you need to specify the required settings in the **Settings** rollout. Select the required reference set from the **Reference Set** drop-down list in the **Settings** rollout.

   In the **Component Origin** drop-down list, the **WCS** option is selected by default. As a result, the absolute coordinate system will be oriented exactly as the coordinate system of the displayed part. If you select the **Absolute** option from the **Component Origin** drop-down list, the absolute coordinate system will be oriented exactly as the coordinate system of the work part.

   In the **Settings** rollout, the **Delete Original Objects** check box is selected by default. As a result, the selected body will be saved as a new part file at the specified location, but it will be deleted from its original location. If you clear this check box, the selected body will be saved as a new part file at the specified location without being deleted from the original location in the part assembly.

6. After setting the required options in the **Create New Component** dialog box, choose the **OK** button; the name and node of the newly created part file will be displayed in the **Assembly Navigator** cascading menu below the parent assembly file node.

7. If you double-click on the name of the newly created part file in the **Assembly Navigator** cascading menu, the part will get activated, and the other parts as well as the parent assembly will get deactivated. You can modify the created part as per your requirement.

8. To activate the parent assembly, double-click on the name of the parent assembly in the **Assembly Navigator** cascading menu.

To create the remaining parts of the parent assembly, you need to follow the same procedure as mentioned above.

9. After creating all the parts of the assembly in the assembly file, choose the **Save** button; all parts will be saved as separate part files at the specified location.

Applying Assembly Constraints

After creating different parts in the Assembly file using the Top-down Assembly design approach and then saving them as separate part files, you need to apply the assembly constraints to the components by using the **Assembly Constraints** tool for making fully constrained assembly.

TUTORIAL

Tutorial 1

In this tutorial, you will create the Radial Engine assembly shown in Figures 9-2 and 9-3. The Radial Engine assembly will be created in two parts, one will be the sub-assembly and the other will be the main assembly. The dimensions of the components of the Radial Engine assembly are shown in Figures 9-4 through 9-8. Save the assembly file with the name *Radial Engine.prt* at the location *\NX\c09\Radial Engine.* **(Expected time: 4 hr)**

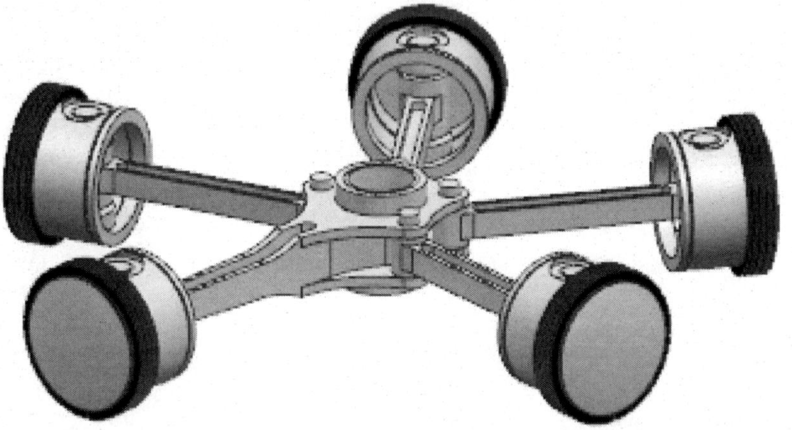

Figure 9-2 *The Radial Engine assembly*

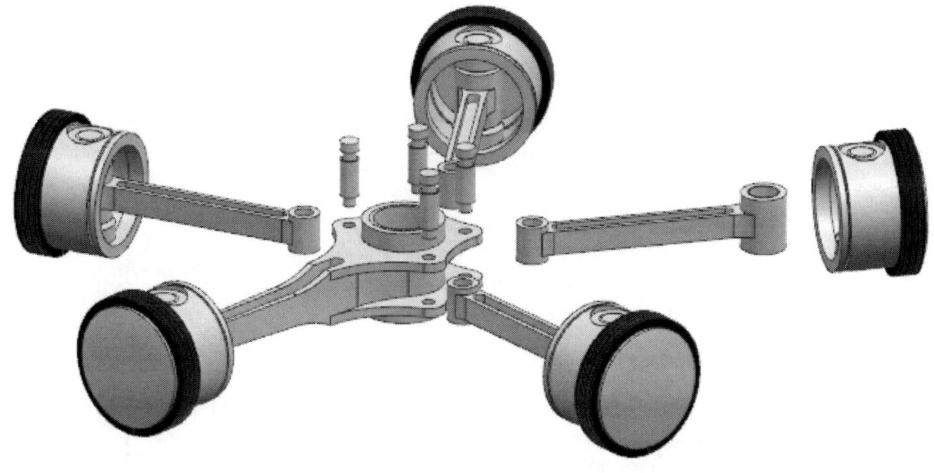

Figure 9-3 *The exploded view of the assembly*

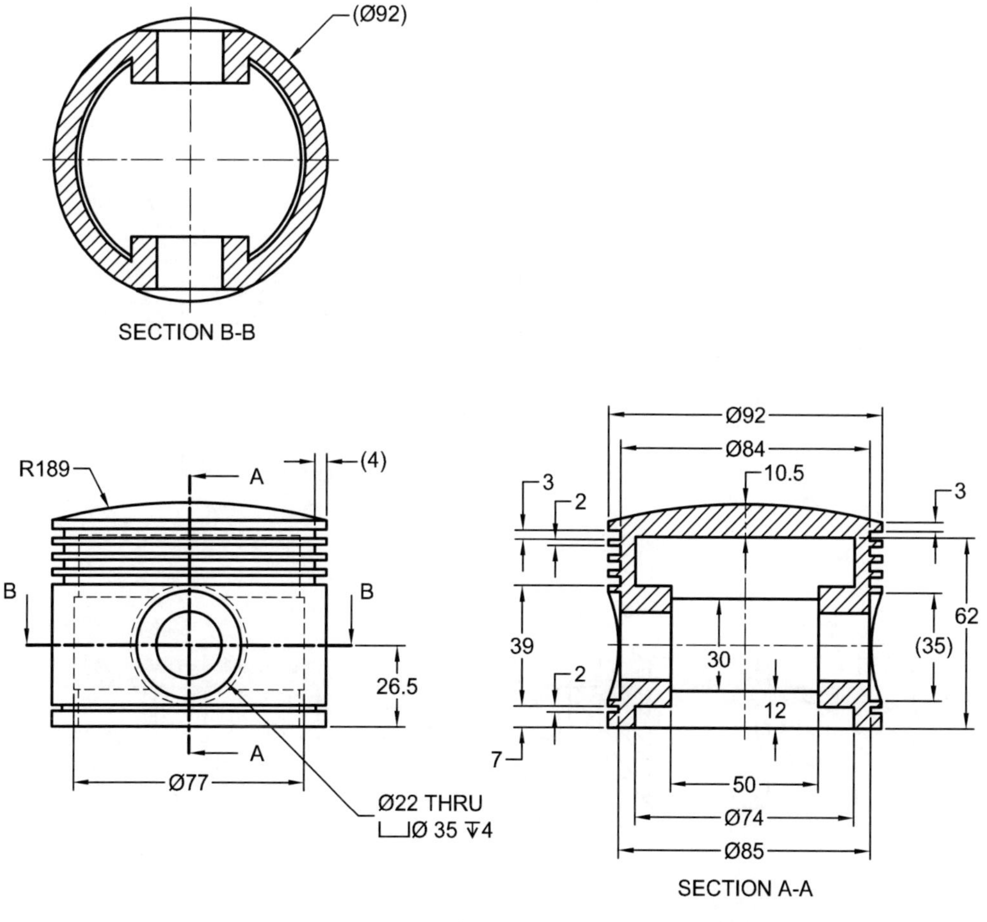

Figure 9-4 *Views and dimensions of the Piston*

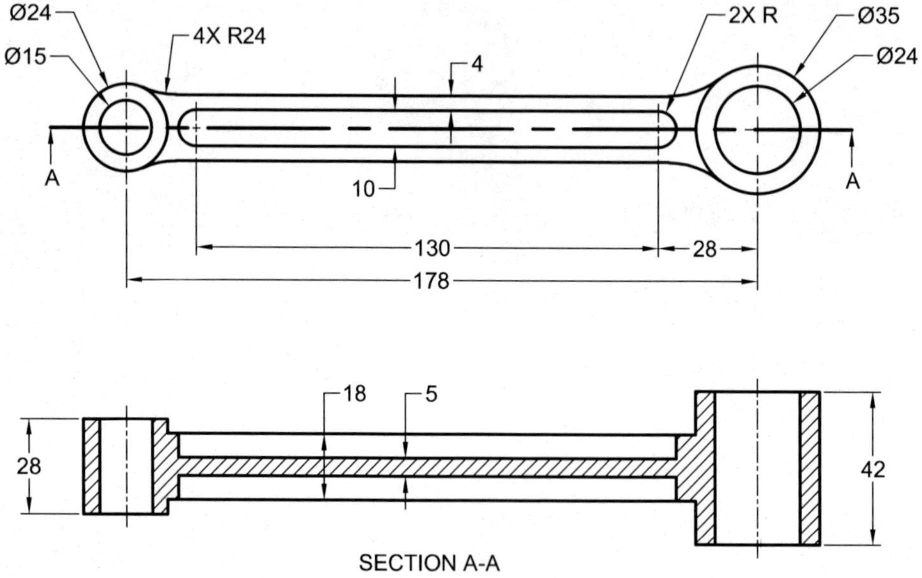

Figure 9-5 Views and dimensions of the Articulated Rod

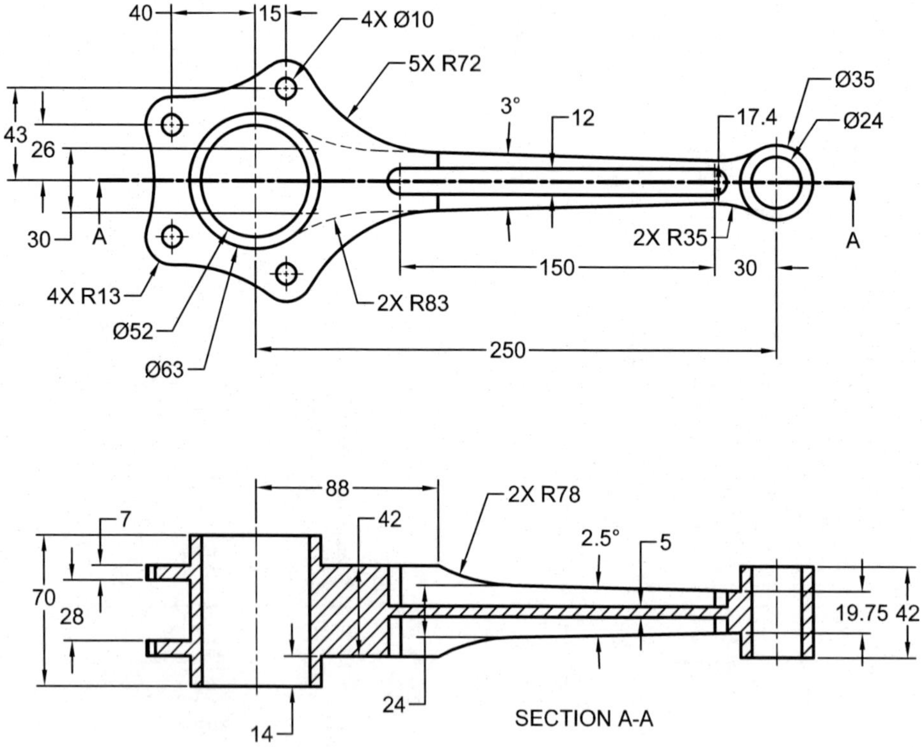

Figure 9-6 Views and dimensions of the Master Rod

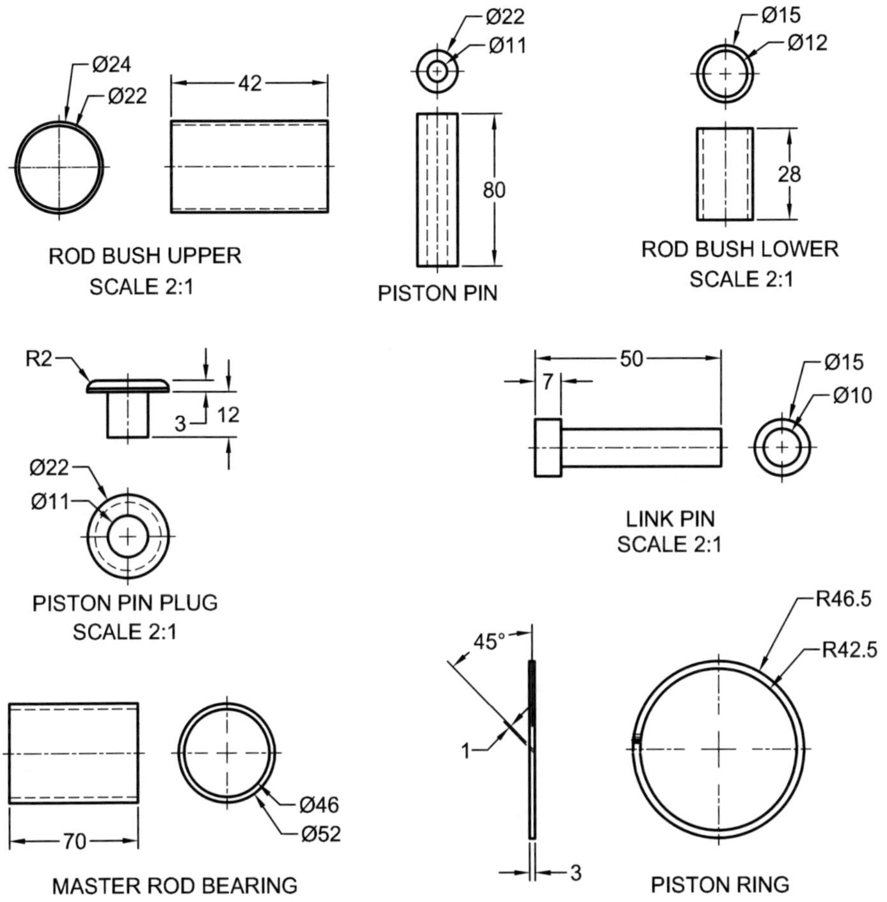

Figure 9-7 *Views and dimensions of the Rod Bush Upper, Rod Bush Lower, Piston Pin Plug, Piston Pin, Piston Rings, Link Pin, and Master Rod Bearing*

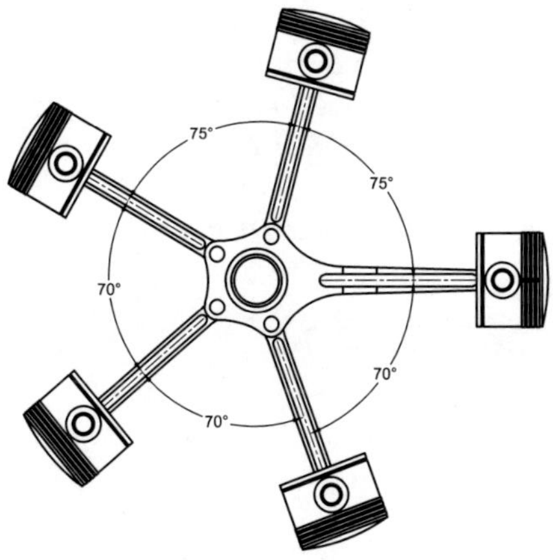

Figure 9-8 *The assembly structure to be followed while assembling the subassemblies with the Master Rod*

Since this is a large assembly, you need to create it in two steps. First, you need to create the sub-assembly that consists of the Articulated Rod, Piston, Rings, Piston Pin, Rod Bush Upper, Rod Bush Lower, and Piston Pin Plug. Next, you will create the main assembly by assembling the Master Rod with the Piston, Piston Rings, Piston Pin, Master Rod Bearing, and Piston Pin Plug, and the instances of the sub-assembly.

The following steps are required to complete this tutorial:

a. Create all the components of the Radial Engine assembly and save them in the *Radial Engine* folder.
b. Start a new file and create the sub-assembly, refer to Figures 9-9 through 9-11.
c. Start a new assembly and then assemble the sub-assembly and the individual components to create the main assembly, refer to Figures 9-14 and 9-15.
d. Save the file.

Creating the Components

1. Create a folder with the name *Radial Engine* at */NX/c09*. Create all the components of the Radial Engine as individual part files and save them in this folder. The views and dimensions of the components are given in Figures 9-4 through 9-8.

Creating the Sub-assembly

1. Start a new file with the name Articulated Rod_Piston and save it in the same folder where the parts were saved. Next, invoke the Assembly environment.

2. Assemble the components in the sequence as illustrated in Figure 9-9. The assembly after assembling the Rod Bush Upper, Rod Bush Lower, Piston, Piston Pin, and Piston Pin Plugs is shown in Figure 9-10.

3. Assemble the Piston Ring to the Piston by using the assembly constraints. The completed Articulated Rod_Piston sub-assembly is shown in Figure 9-11. Next, choose the **Save** button.

Note

*1. You can change the color of the Piston Rings for a better display. To do so, choose **Menu > Edit > Object Display** from the **Top Border Bar**; the **Class Selection** dialog box is displayed and you are prompted to select the objects to edit. Select the Piston Rings from the sub-assembly and choose the **OK** button; the **Edit Object Display** dialog box is displayed. Choose the **Color** swatch; the **Color** dialog box is displayed. Select the required color and then choose the **OK** button from the **Color** dialog box; the **Edit Object Display** dialog box is displayed again. Next, choose the **OK** button from it.*

*2. In Figure 9-10, the solid texture of the Piston has been changed to the transparent texture. This is done to view the components that are assembled inside the Piston. To change the texture of the Piston from solid to transparent, select the Piston from the drawing window and then invoke the **Edit Object Display** dialog box. Next, move the **Translucency** sliding bar to the value of 75 and then choose the **OK** button from the **Edit Object Display** dialog box; the warning message displayed. Choose the **OK** button to close the dialog box. Note that the transparency value of the object ranges from 0 to 100.*

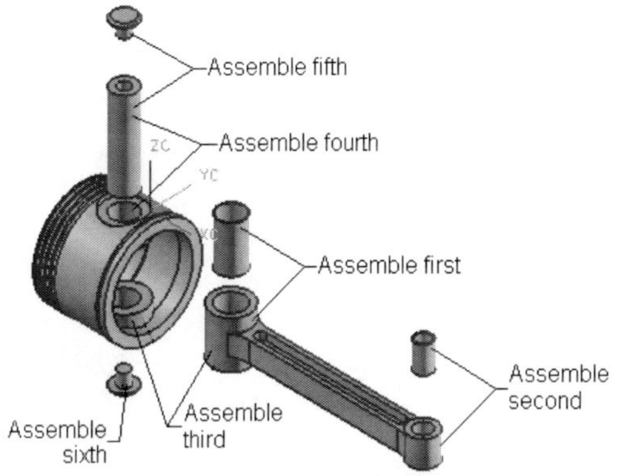

Figure 9-9 Sequence for assembling the components to create the sub-assembly

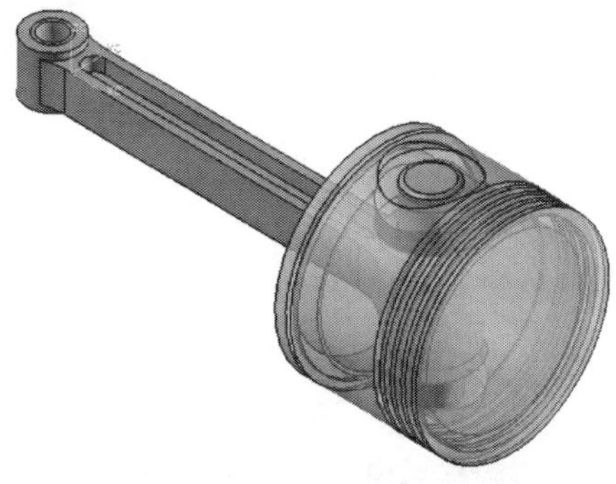

Figure 9-10 The sub-assembly after assembling the Rod Bush Lower, Rod Bush Upper, Articulated Rod, Piston, Piston Pin, and Piston Pin Plug

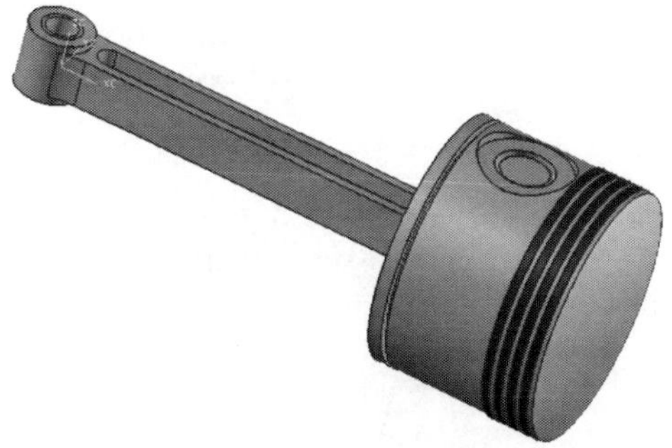

Figure 9-11 The sub-assembly after assembling all instances of the Piston Rings

Creating the Main Assembly

In this section, you need to create the main assembly by assembling the sub-assembly and the individual components.

1. Start a new file with the name Radial Engine and specify its location in the same folder where the components and the sub-assembly are saved. Next, invoke the Assembly environment.

2. Assemble the Master Rod, Rod Bush Upper, and Master Rod Bearing using the assembly constraints, refer to Figure 9-12.

3. Next, assemble the Piston, Piston Pin, and Piston Pin Plugs using the assembly constraints, refer to Figure 9-12.

4. Assemble the Piston Rings with the Piston. Change the color of the Piston Rings. The assembly after assembling these components is shown in Figure 9-12.

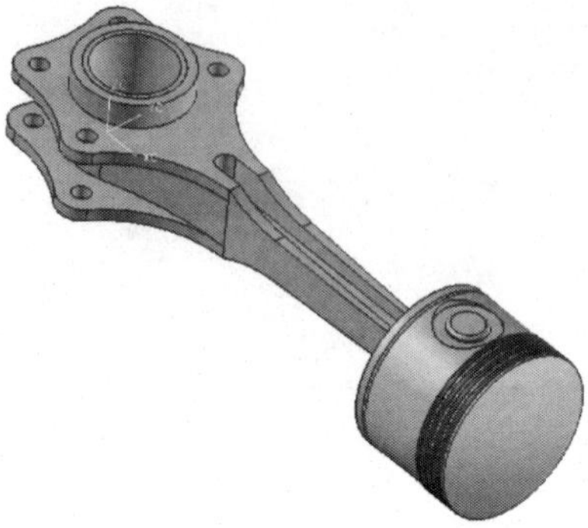

Figure 9-12 Components assembled in the main assembly

Assembling the Sub-assembly with the Main Assembly

Next, you need to assemble the instances of the Articulated Rod_Piston sub-assembly with the main assembly by using the assembly constraints.

1. Choose the **Add** tool from the **Assemblies** tab; the **Add Component** dialog box is displayed.

2. Choose the **Open** button from the **Add Component** dialog box; the **Part Name** dialog box is displayed. Select the Articulated Rod_Piston sub-assembly from the dialog box and choose the **OK** button.

3. Assemble the first instance of the Articulated Rod_Piston sub-assembly with the main assembly by using the assembly constraints. The assembly after assembling the first instance of the sub-assembly is shown in Figure 9-13.

Figure 9-13 The main assembly after assembling the first instance of the sub-assembly

4. Similarly, assemble other instances of the sub-assembly with the main assembly by using the assembly constraints. The assembly structure to be followed while assembling the subassemblies with the main assembly is shown in Figure 9-8.

The assembly after assembling all the instances of the sub-assembly is shown in Figure 9-14.

Assembling the Link Pin

1. Assemble four instances of the Link Pin with the Master Rod by using the assembly constraints. Modify the color of the Link Pin. The resulting assembly after assembling the Link Pins is shown in Figure 9-15.

Figure 9-14 *The assembly after assembling all instances of the sub-assembly with the main assembly*

Figure 9-15 *The final Radial Engine assembly*

2. Choose **Menu > File > Save** from the **Top Border Bar** to save the assembly. Next, close the file.

EXERCISE

Exercise 1

Create the Shaper Tool Head assembly shown in Figure 9-16. After creating the assembly, create its exploded view, as shown in Figure 9-17. The dimensions of the components are given in Figures 9-18 through 9-22. Save the assembly file with the name *Shaper.prt* at the location *\NX\c09\Shaper Tool head.* **(Expected time: 4 hr)**

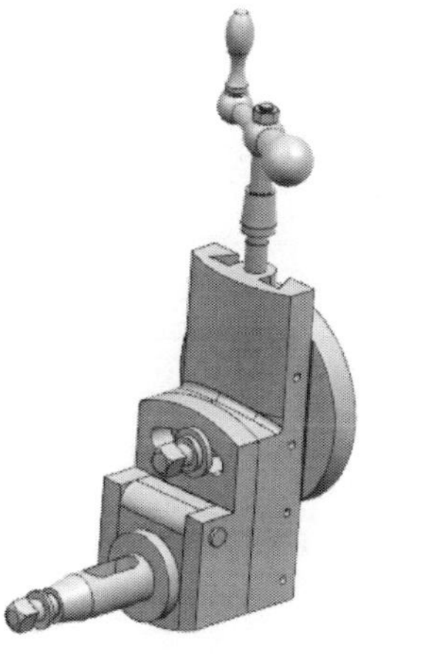

Figure 9-16 *The Shaper Tool Head assembly*

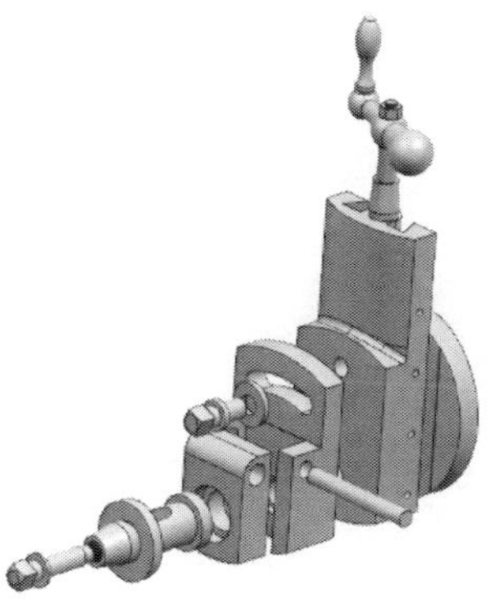

Figure 9-17 *The exploded view of the Shaper Tool Head assembly*

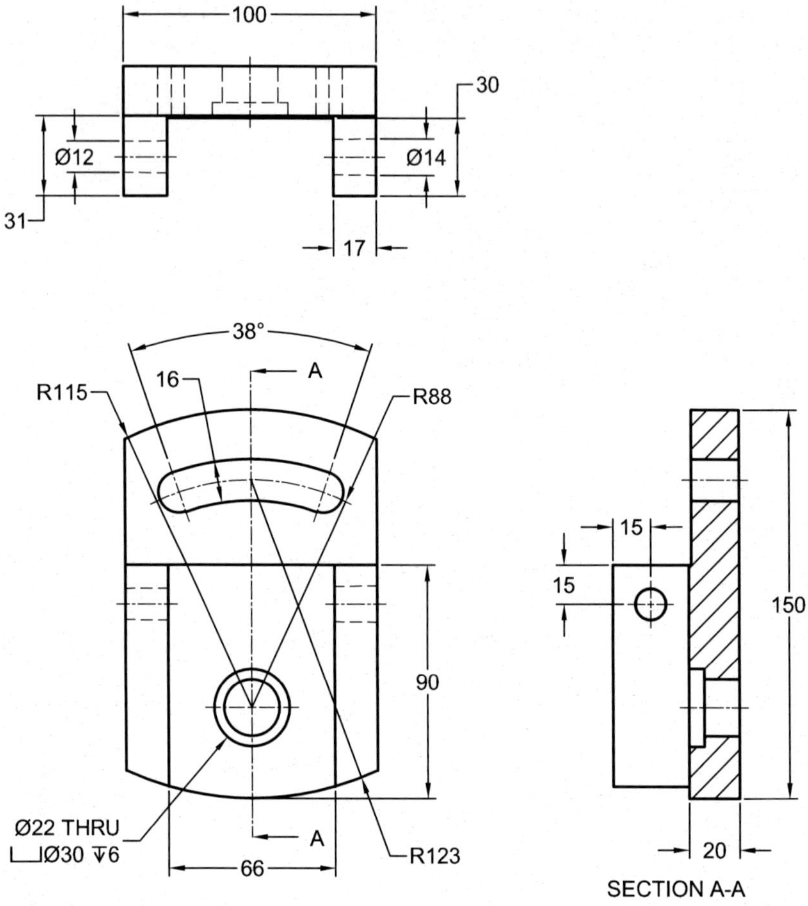

Figure 9-18 *Views and dimensions of the Swivel Plate*

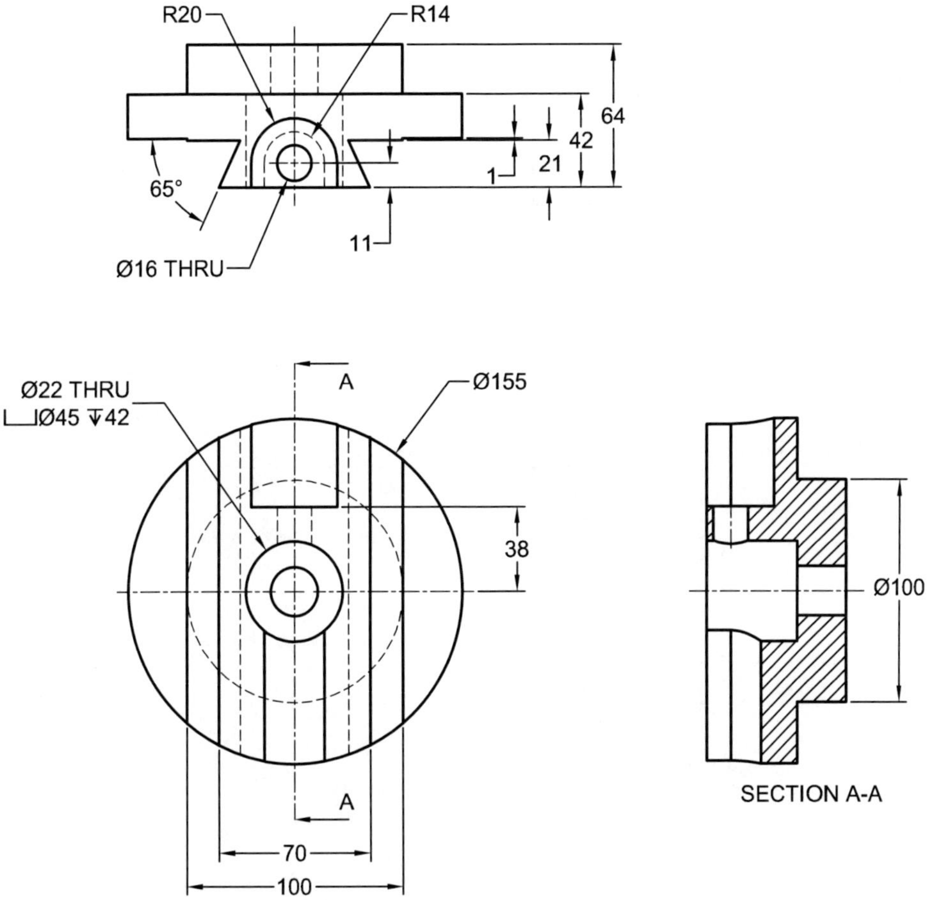

Figure 9-19 *Views and dimensions of the Back Plate*

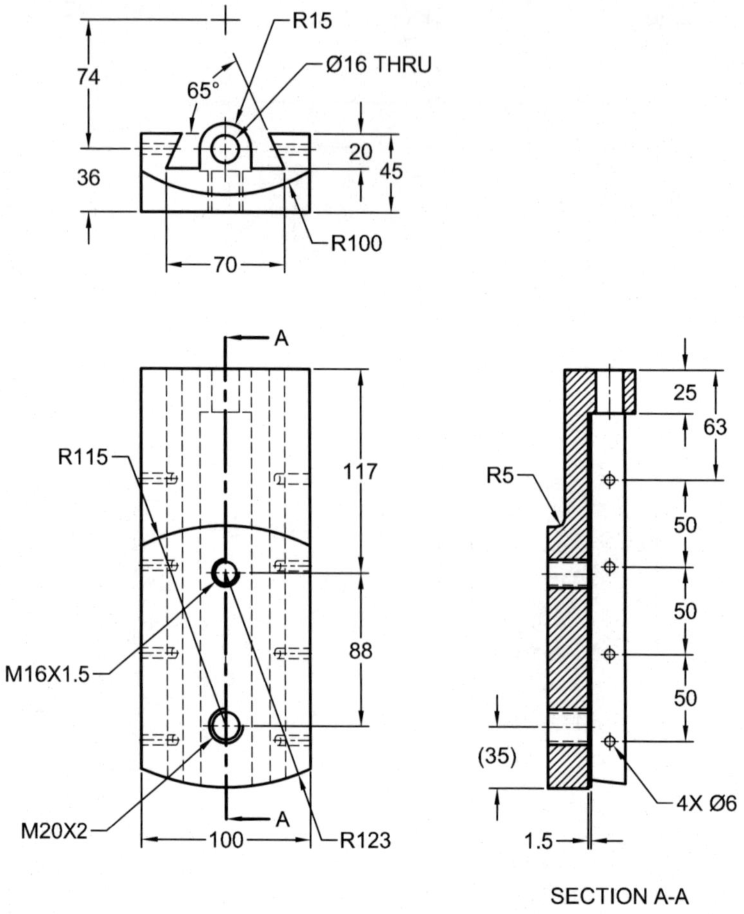

Figure 9-20 *Views and dimensions of the Vertical Slide*

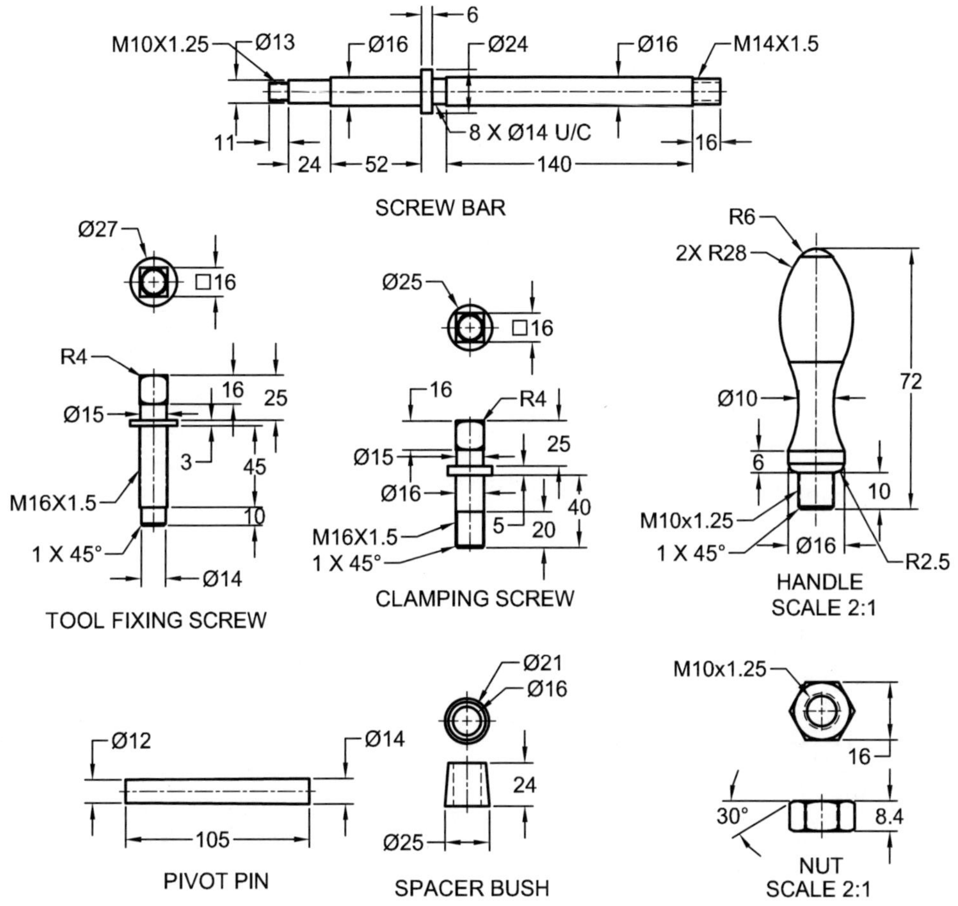

Figure 9-21 *Views and dimensions of various components*

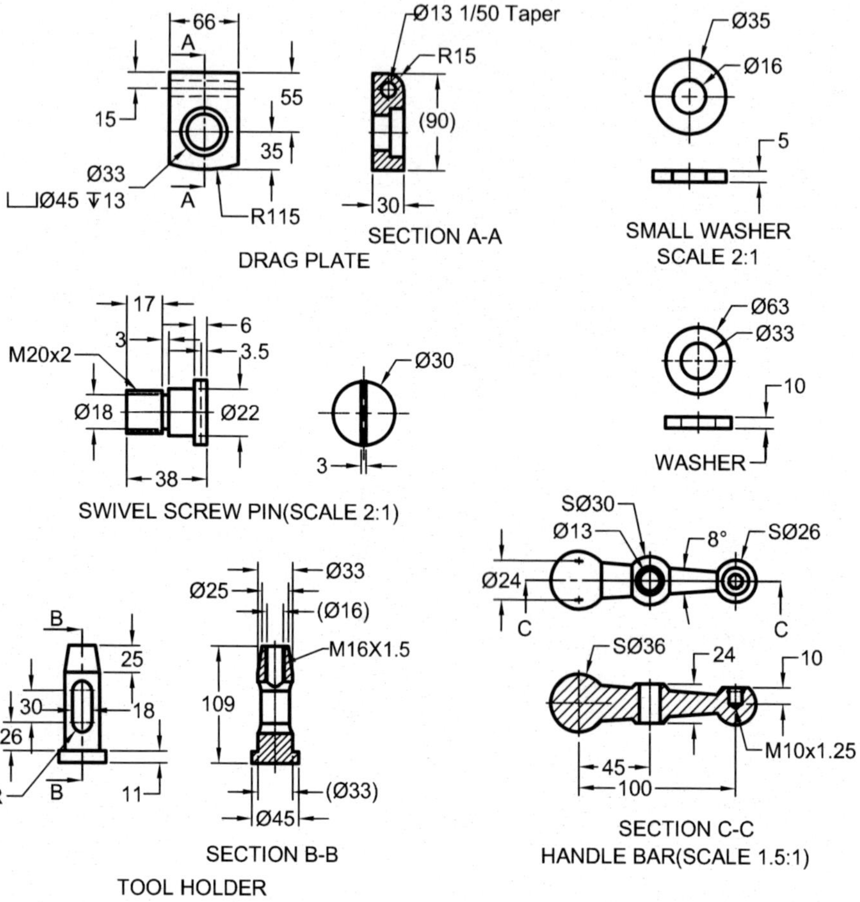

Figure 9-22 Views and dimensions of various components

Chapter 10

Surface Modeling

Learning Objectives

After completing this chapter, you will be able to:

- *Create extrude surfaces*
- *Create revolve surfaces*
- *Create combined surfaces*
- *Create surfaces using the Through Curves*
- *Add thickness to a surface*

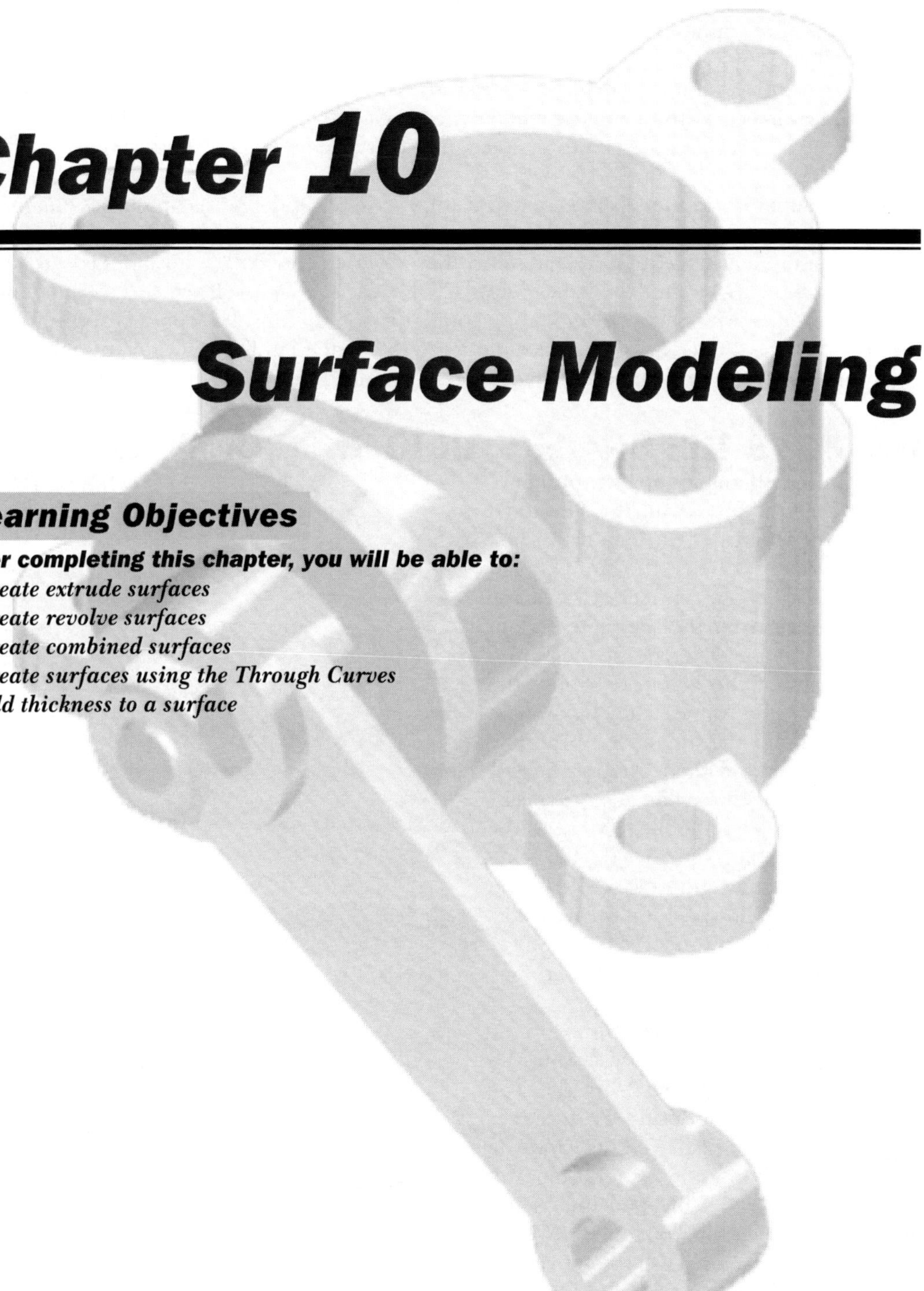

INTRODUCTION TO SURFACE MODELING

Surfaces are three dimensional (3D) bodies with negligible thickness. They are used extensively for modeling complex features. A model or an assembly created using the surface body type has surface area but no volume or mass properties. In NX, surfaces are created in the form of single or multiple patches. With the increase in patches, the control over the shape of the surface also increases. In NX, surfaces are known as sheets and surface modeling is known as sheet modeling.

Most of the real world models are created using the solid modeling techniques. Only models that are complex in shape and have a non-uniform surface area are created with the help of the surface modeling techniques. The tools that are used to create solid models can also be used to create surface models. It becomes easy for the readers to learn surface modeling if they are familiar with the solid modeling tools. In NX, there is no separate application for surfaces. You need to create the surface model in the Shape Studio environment or Modeling environment. Before creating the surface model, you need to change the body type to sheet.

INVOKING THE SHAPE STUDIO ENVIRONMENT

To invoke the Shape Studio environment, choose theNew tool from the **Standard** group of the **Home** tab; the **New** dialog box will be displayed. Choose the **Model** tab, if it is not already chosen and select the **Shape Studio** template from the **Templates** rollout of the dialog box. Next, choose the button on the right of the **Name** text box; the **Choose New File Name** dialog box will be displayed. Enter the file name in the **File name** edit box and then, choose the **OK** button to exit the dialog box. Also, to specify the location to save the file, browse to the folder where you need to save the file using the button on the right of the **Folder** edit box. After specifying the location of the file, choose the **OK** button twice; the Shape Studio environment will be displayed.

*Figure 10-1 The **Modeling Preferences** dialog box*

After invoking the Shape Studio or the Modeling environment, choose **Menu > Preferences > Modeling** from the **Top Border Bar**; the **Modeling Preferences** dialog box will be displayed, as shown in Figure 10-1. Choose the **General** tab, if it is not chosen by default, and then select the **Sheet** radio button from the **Body Type** area. Next, choose the **OK** button to exit this dialog box. All models created, henceforth, in the Modeling environment will be the sheet models.

Creating an Extruded Surface

Ribbon: Home > More Gallery > Design Feature > Extrude
Menu: Insert > Design Feature > Extrude

As mentioned earlier, there is no separate tool available for creating an extruded surface. After invoking the Shape Studio environment, you can use the **Extrude** tool to create extruded sheets. The sketch drawn for creating an extruded surface may be an open or a closed entity. After creating the sketch, choose the **Extrude** tool from **Menu > Insert > Design Feature** in the **Top Border Bar**; the **Extrude** dialog box will be displayed and you will be prompted to select the section geometry to extrude. Select the sketch and enter the start and end extrusion values in their respective **Distance** edit boxes available below the **Start** and **End** drop-down lists of the **Limits** rollout in the dialog box. Next, choose the **OK** button from the **Extrude** dialog box; a sheet will be created. The options in the **Extrude** dialog box are the same as those discussed in Chapter 4. Figures 10-2 and 10-3 show the extruded surfaces created by using the open and closed sketches, respectively.

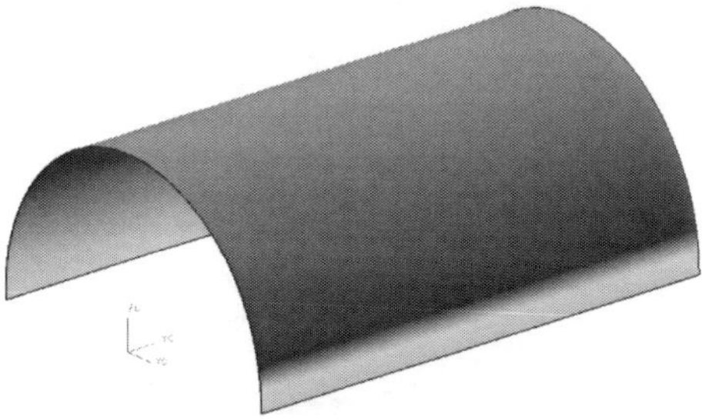

Figure 10-2 *Extruded surface created on an open sketch*

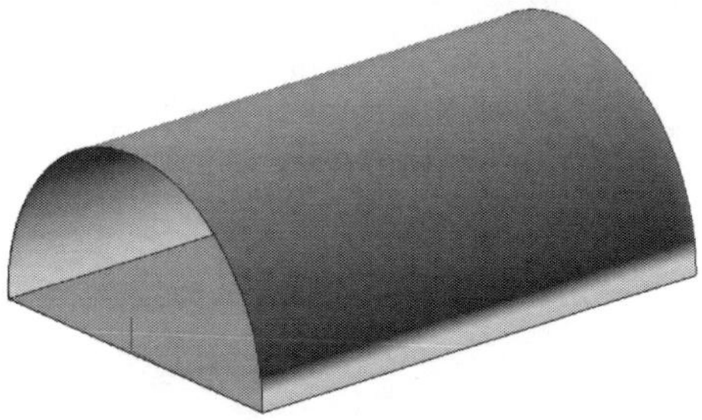

Figure 10-3 *Extruded surface created on a closed sketch*

Note

*You can use only the **None** and **Inferred** options from the **Boolean** drop-down list in the Shape Studio environment. The other options of this drop-down list are not available in this environment.*

Creating a Revolved Surface

Ribbon: Home > More Gallery > Design Feature > Revolve
Menu: Insert > Design Feature > Revolve

The **Revolve** tool is used to create a revolved surface. To create a revolved surface, first create a sketch, and then choose the **Revolve** tool from **Menu > Insert > Design Feature** in the **Top Border Bar**; the **Revolve** dialog box will be displayed. Also, you will be prompted to select the section geometry. Select the sketch drawn for the revolved surface. Next, click on the **Specify Vector** area in the **Axis** rollout of the dialog box and specify the axis of revolution; you will be prompted to select object to infer point. Specify the point where you want to locate the vector and then specify the start and end angles in the **Angle** edit boxes. Next, choose the **OK** button; a revolved surface will be created. The revolved surface models created by using an open sketch and a closed sketch are shown in Figures 10-4 and 10-5, respectively.

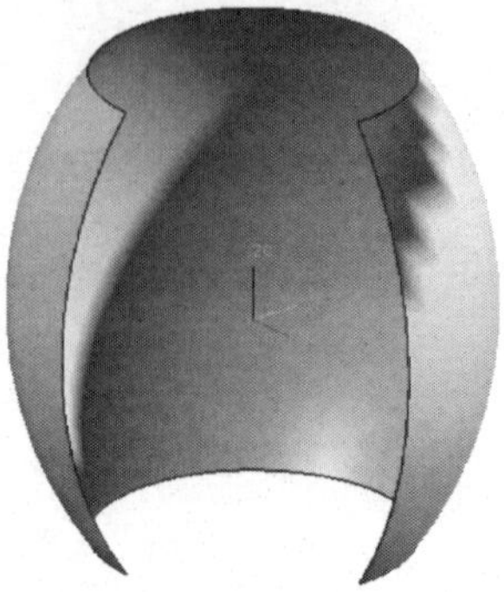

Figure 10-4 Revolved surface created using an open sketch

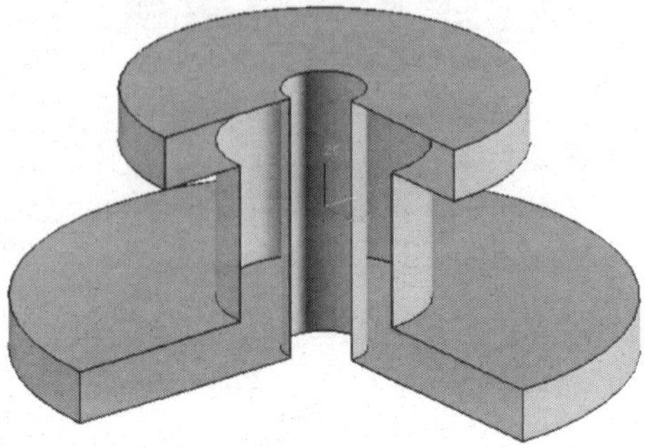

Figure 10-5 Revolved surface created using a closed sketch

Creating a Combined Surface

Menu: Insert > Combine > Combine

You can use the **Combine** tool to trim and join regions of multiple sheet bodies. To do so, choose **Menu > Insert > Combine** from the **Top Border Bar**; the **Combine** dialog box will be displayed, refer to Figure 10-6.

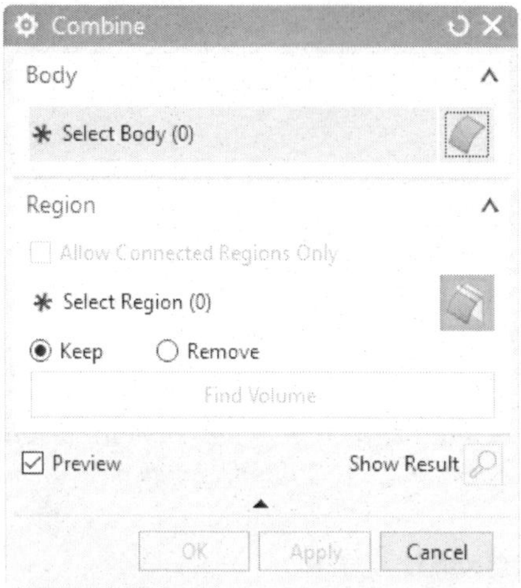

*Figure 10-6 The **Combine** dialog box*

The rollouts available in the **Combine** dialog box are discussed next.

Body Rollout

The **Body** button in this rollout is activated by default. As a result, you are prompted to select the bodies to be combined. Select the bodies to be combined using this rollout.

Region Rollout

The options available in this rollout are discussed next.

Allow Connected Regions Only

If this check box is selected, you can only select the bodies that are edge connected to a previously selected body. On clearing this check box, you can select any body created during the session.

Select Region

This area allows you to select the region of a body to keep or remove. If you select the **Keep** radio button, the selected side of the bodies will be retained and if you select the **Remove** radio button; the selected side of the bodies will be trimmed.

Find Volume

This button allows you to find and select regions that enclose a solid volume, or connected regions that form a volume-like sheet.

Figure 10-7 shows sheet bodies selected to be combined. Figure 10-8(a) shows the resulting body when the **Keep** radio button is selected and Figure 10-8(b) shows the resulting body when the **Remove** radio button is selected.

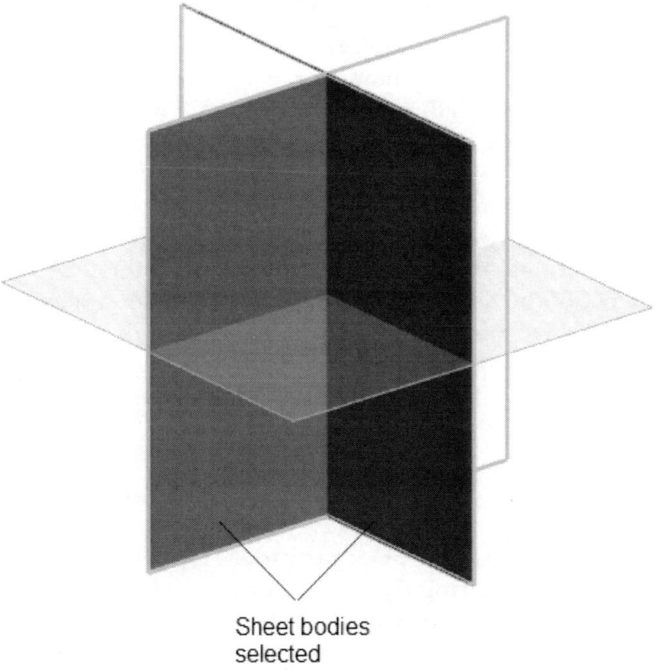

Figure 10-7 *Sheet bodies selected to be combined*

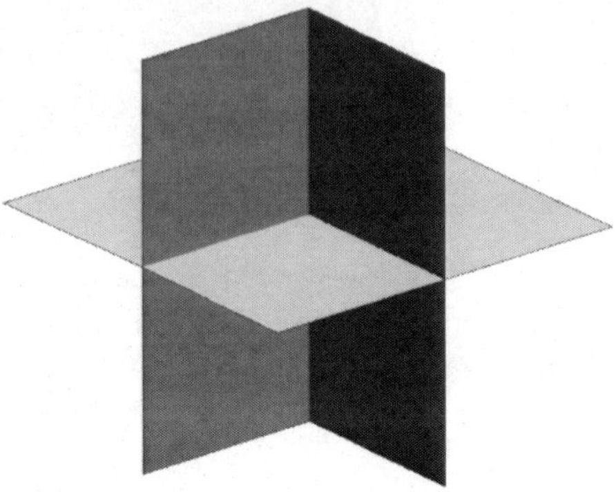

Figure 10-8(a) *Resulting sheet body when the **Keep** radio button is selected*

Figure 10-8(b) *Resulting sheet body when the **Remove** radio button is selected*

Creating a Surface Using the Through Curves Tool

Ribbon:	Home > More Gallery > Mesh Surface > Through Curves
Menu:	Insert > Mesh Surface > Through Curves

You can create surfaces with multiple section strings using the **Through Curves** tool. This method allows you to select any number of section strings. To do so, choose the **Through Curves** tool from the **Mesh Surface** gallery of the **More** gallery, refer to Figure 10-9; the **Through Curves** dialog box will be displayed, as shown in Figure 10-10, and you will be prompted to select the curve or point to section. Select the strings for first section and press the middle mouse button; you will be prompted again to select the curve or point to section. Likewise, you can select any number of section strings. After selecting section strings, make sure that all the arrows on the section strings point in the same direction. All the selected sections will be listed in the **List** sub-rollout of the **Sections** rollout. You can reorder a selected section by using the **Move Up** and **Move Down** buttons available on the right of the **List** sub-rollout. You can also delete a selected section by using the **Remove** button.

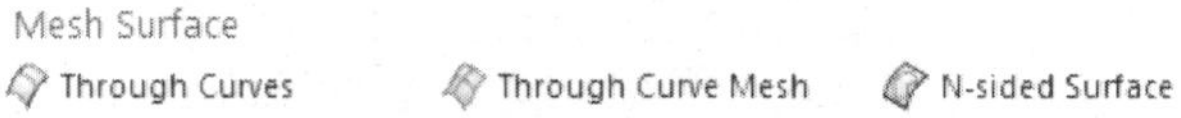

Figure 10-9 *The **Mesh Surface** gallery of the **More** gallery*

In the **Patch Type** drop-down list of the **Output Surface Options** rollout, there are three options, **Single**, **Multiple**, and **Match String**. If you select the **Single** option, a surface will be created with a single patch. If you select the **Multiple** option, the surface will be created with multiple patches. The number of patches formed depends upon the **Alignment** option selected from the **Alignment** rollout.

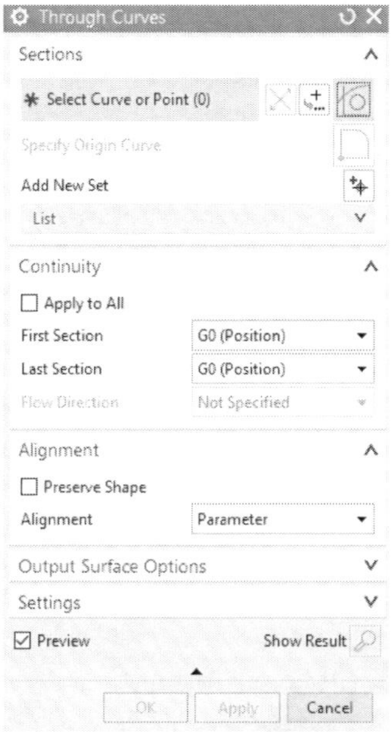

Figure 10-10 *The* ***Through Curves*** *dialog box*

When you select the **Multiple** option from the **Patch Type** drop-down list of the **Output Surface Options** rollout, the **Closed in V** and **Normal to End Sections** check boxes will be enabled. If you select the **Closed in V** check box, the surface body will be closed in the V direction and the **Normal to End Sections** check box will be deactivated. If you select the **Normal to End Sections** check box, the resultant surface will be normal to the two end sections and the options in the **Continuity** rollout will be deactivated. Figure 10-11 shows the section strings selected for creating a through curve surface and Figure 10-12 shows the resulting surface.

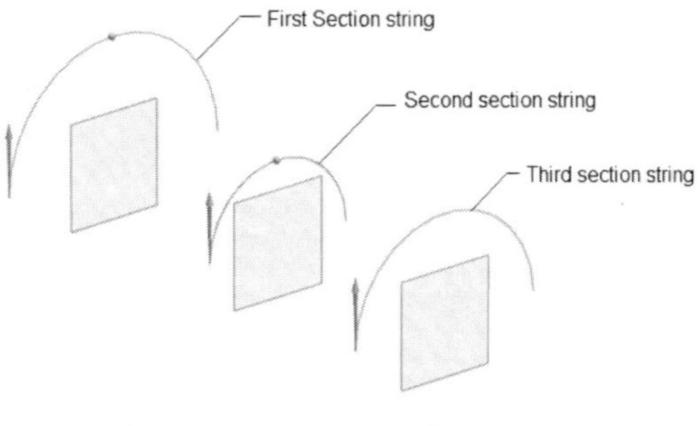

Figure 10-11 *Section strings selected for creating the Through Curves surface*

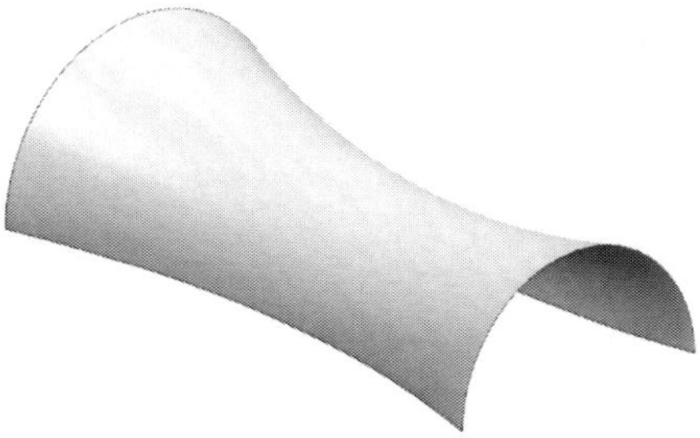

Figure 10-12 *The resulting surface*

Adding Thickness to a Surface

Ribbon:	Home > More Gallery > Offset/Scale > Thicken (Customize to add)
Menu:	Insert > Offset/Scale > Thicken

The **Thicken** tool is used to add thickness to a sheet. Once you add thickness to a sheet, it is converted into a solid. To add thickness to a sheet, choose **Menu > Insert > Offset/Scale > Thicken** from the **Top Border Bar**; the **Thicken** dialog box will be displayed, as shown in Figure 10-13, and you will be prompted to select faces to thicken. Select the sheet bodies to which the material is to be added.

Enter thickness value in the **Offset 2** edit box to add thickness along the direction of handle. Enter a negative thickness value in the **Offset 1** edit box to assign thickness on the other side of the surface.

The **Boolean** drop-down list in the **Boolean** rollout provides options to perform the boolean operations on an existing solid body. Select the required option from the **Boolean** drop-down list and choose the **Body** button from the **Boolean** rollout. Next, select the target solid body; the boolean operation will be performed.

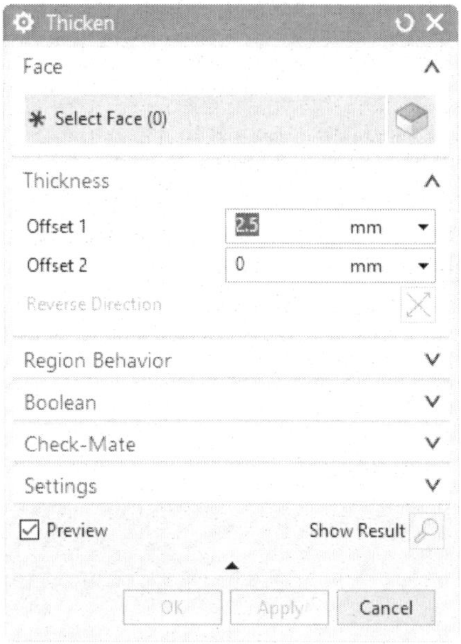

***Figure 10-13** The **Thicken** dialog box*

Figure 10-14 shows the sheet selected for adding thickness and Figure 10-15 shows the resulting solid body.

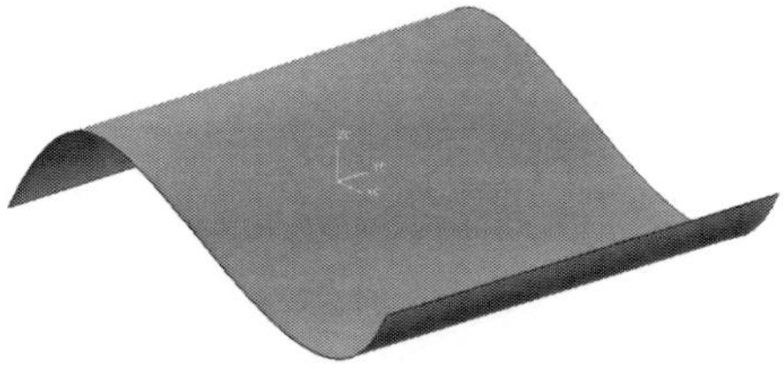

Figure 10-14 *The sheet selected for adding material*

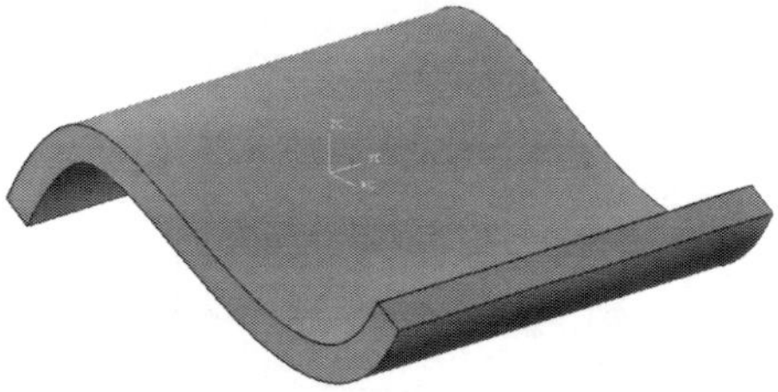

Figure 10-15 *The sheet after adding thickness*

TUTORIAL

Tutorial 1

In this tutorial, you will create the surface model shown in Figure 10-16. The dimensions and orthographic views of the model are shown in Figure 10-17. After creating the surface, save it with the name *c10tut1.prt* at the location *\NX\c10.* **(Expected time: 30 min)**

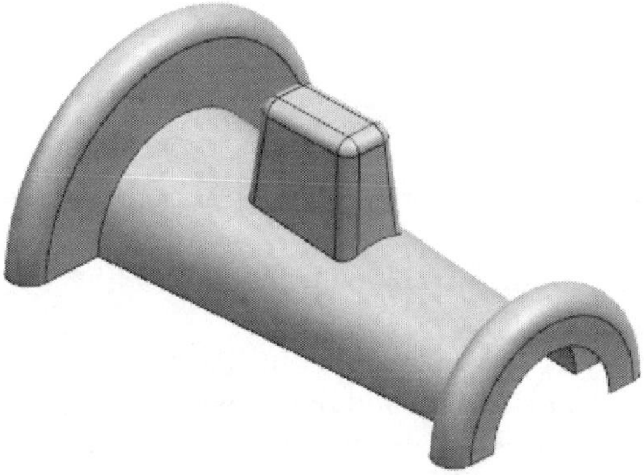

Figure 10-16 *The isometric view of the surface model*

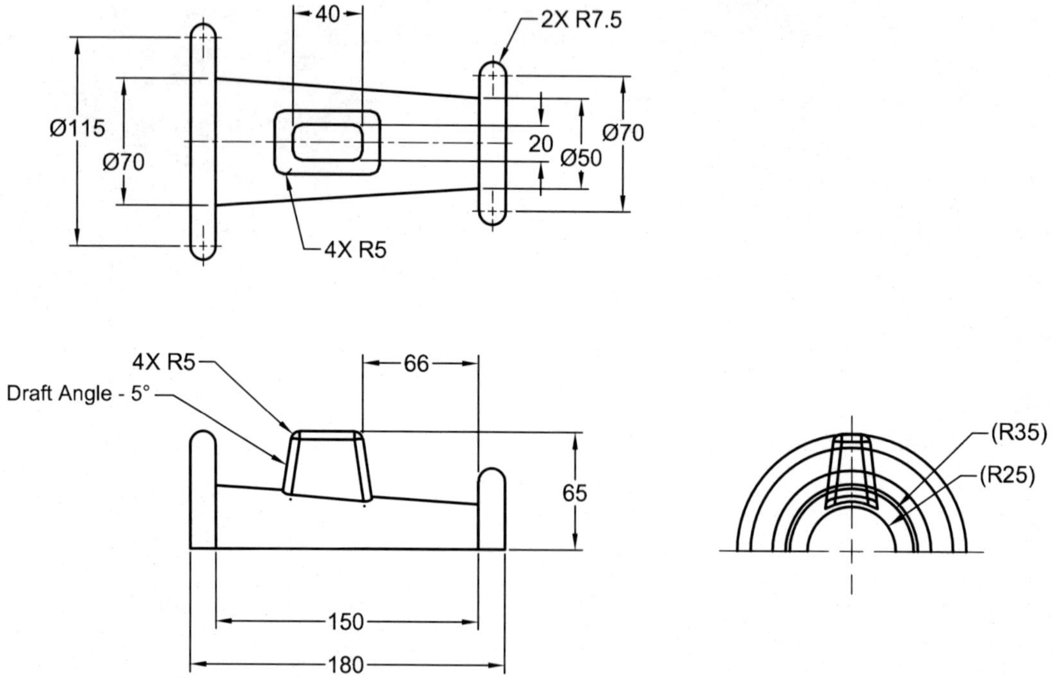

Figure 10-17 *Dimensions and drawing views of the surface model*

The following steps are required to complete this tutorial:

a. Start a new file and then set the sheet modeling environment.
b. Create the sketch for the base surface and then revolve it, refer to Figures 10-18 and 10-19.
c. Create the second feature, which is an extruded surface, refer to Figures 10-20 and 10-21.
d. Trim the base surface, refer to Figures 10-22 and 10-23.
e. Trim the extended part of the third feature, refer to Figures 10-24 and 10-25.
f. Create the top surface of the extruded feature, refer to Figures 10-26 and 10-27.
g. Stitch the bounded plane and extruded surfaces with the revolved surface.
h. Fillet the stitched surface, refer to Figures 10-28.
i. Save and close the file.

Starting a New File and Setting the Sheet Environment

1. Start a new file with the name *c10tut1.prt* using the **Shape Studio** template and specify its location at *C:\NX\c10*.

2. Choose **Menu > Preferences > Modeling** from the **Top Border Bar**; the **Modeling Preferences** dialog box is displayed.

3. In this dialog box, select the **Sheet** radio button from the **Body Type** rollout and choose the **OK** button.

Creating the Base Feature by Revolving the Sketch

1. Create the sketch for the base surface on the XC-YC plane, as shown in Figure 10-18.

2. Revolve the sketch about an angle of 180 degrees. The resulting revolved base feature created by using the **Sheet** option is shown in Figure 10-19.

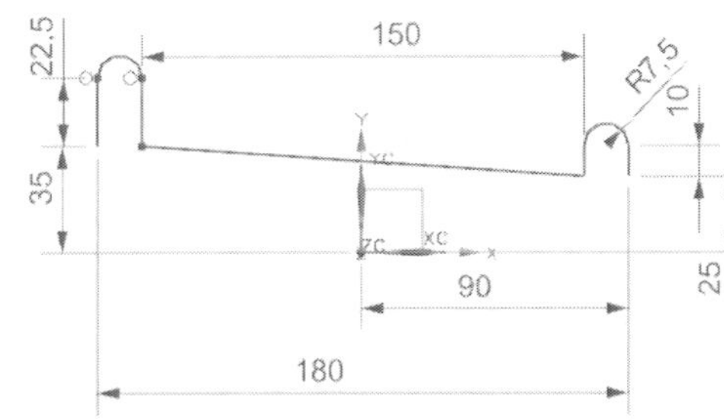

Figure 10-18 The sketch created for the base feature

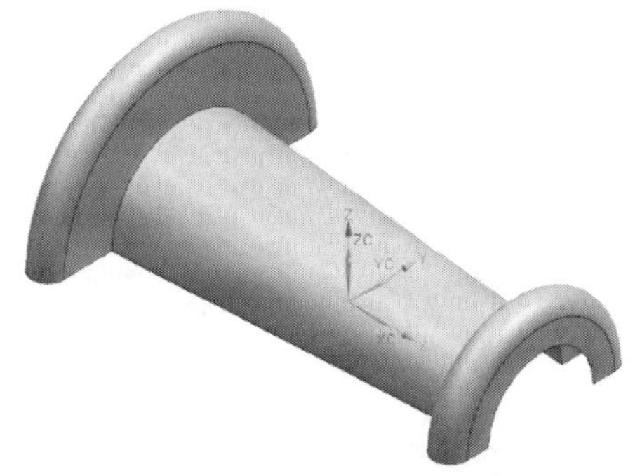

Figure 10-19 The base feature created by revolving the sketch

Creating the Second Feature by Extruding the Sketch

1. Create a datum plane in the upward direction at an offset of 65 mm from the XC-YC plane.

2. Create the sketch for the second feature by selecting the offset plane as the sketching plane, as shown in Figure 10-20.

3. Extrude the sketch through a distance of 50 mm in the downward direction and at the draft angle of -5 degrees by selecting the **From Start Limit** option from the **Draft** drop-down list. The resulting extruded surface model is shown in Figure 10-21.

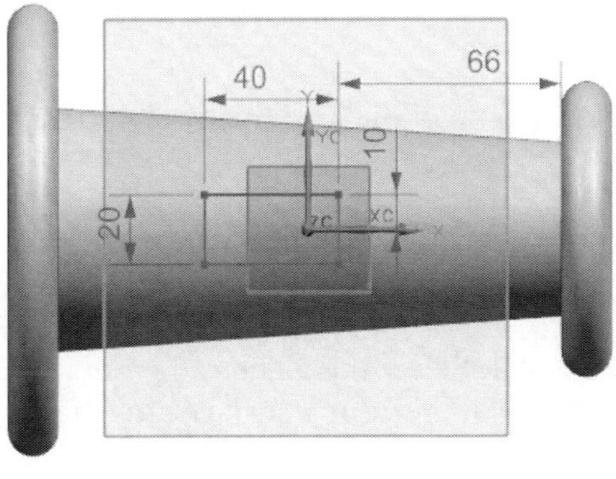

Figure 10-20 The sketch drawn for creating the extruded feature

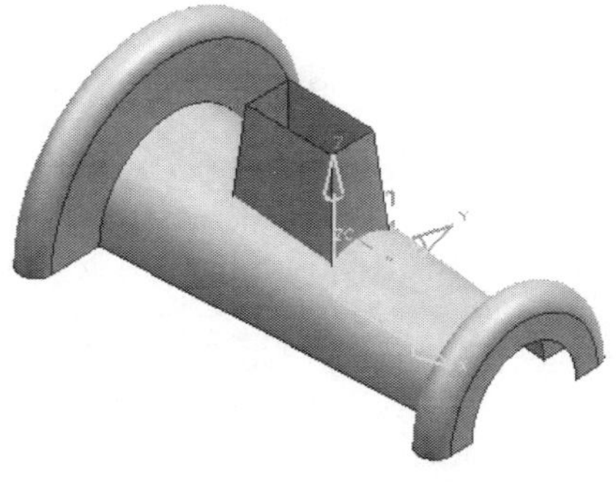

Figure 10-21 The resulting extruded surface

Trimming the Base Surface with Respect to the Second Surface

Next, you need to trim the base surface with respect to the second feature.

1. Choose the **Trim Sheet** tool from **Menu > Insert > Trim** in the **Top Border Bar**; the **Trim Sheet** dialog box is displayed and you are prompted to select the sheet bodies to trim.

2. Select the sheet to be trimmed, as shown in Figure 10-22. Make sure that you select the sheet by using the selection points shown in this figure. After selecting the target sheet body, press the middle mouse button; you are prompted to select the boundary objects.

3. Select the trimming surfaces, as shown in Figure 10-22. Make sure that you select the surfaces by using the points shown in this figure.

4. Select the **Keep** radio button from the **Region** rollout and then choose the **OK** button. The trimmed surface model is shown in Figure 10-23.

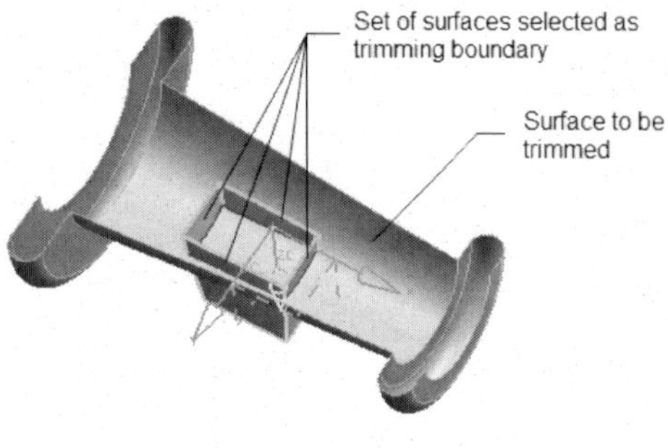

Figure 10-22 *The surface to be trimmed and the trimming surfaces*

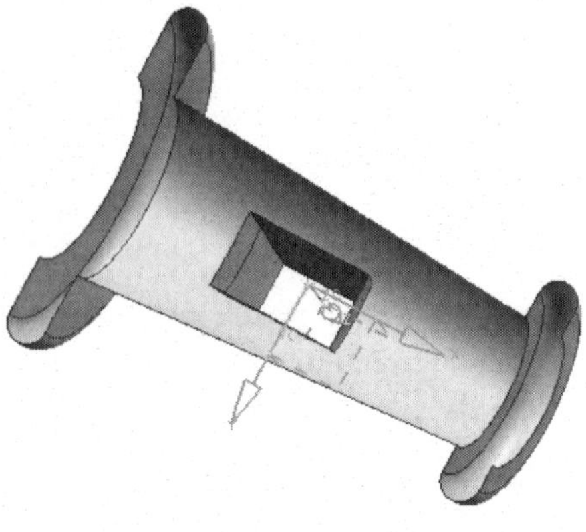

Figure 10-23 *The surface model after trimming the base surface with respect to the extruded surface*

Trimming the Second Surface with the Base Surface

After trimming the base surface, you need to trim the unwanted portions of the second feature.

1. Choose the **Trim Sheet** tool from **Menu > Insert > Trim** of the **Top Border Bar**; the **Trim Sheet** dialog box is displayed and you are prompted to select the sheet bodies to trim.

2. Select the sheet to be trimmed, as shown in Figure 10-24, and then press the middle mouse button; you are prompted to select the boundary objects.

3. Select the trimming surface, as shown in Figure 10-24. Select the **Keep** radio button from the **Region** rollout and then choose the **OK** button. The resulting surface after trimming the extended portion of the extruded surface is shown in Figure 10-25.

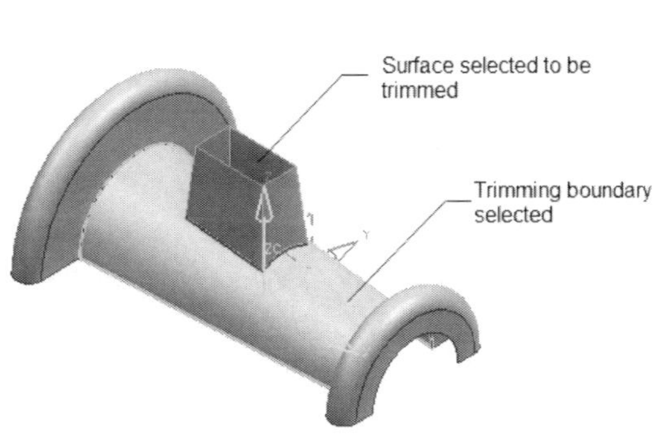

Figure 10-24 The surface to be trimmed and the trimming surface

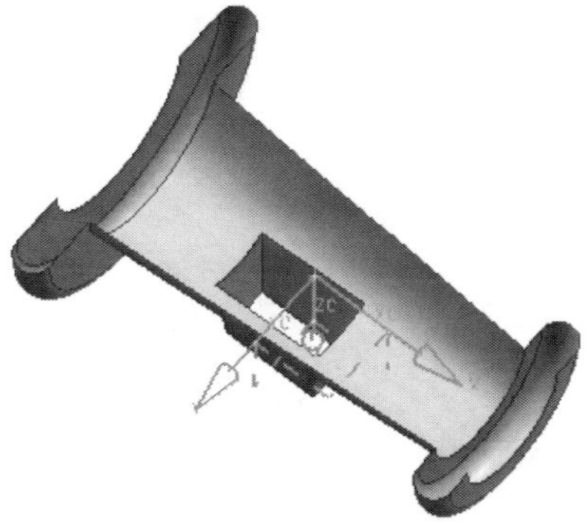

Figure 10-25 The resulting surface after trimming the extended portion of the extruded surface

Creating the Top Surface on the Extruded Feature

Next, you need to create the top surface of the extruded feature.

1. Choose the **Four Point Surface** tool from the **Surface Drop-down** in the **Create** group; the **Four Point Surface** dialog box is displayed and you are prompted to select object to infer point.

2. Select the vertices at the top of the extruded feature, refer to Figure 10-26, and then choose the **OK** button. The resulting surface is shown in Figure 10-27.

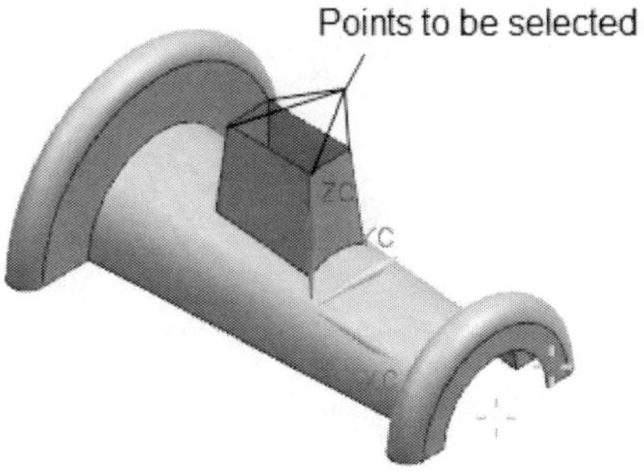

Figure 10-26 Vertices to be selected

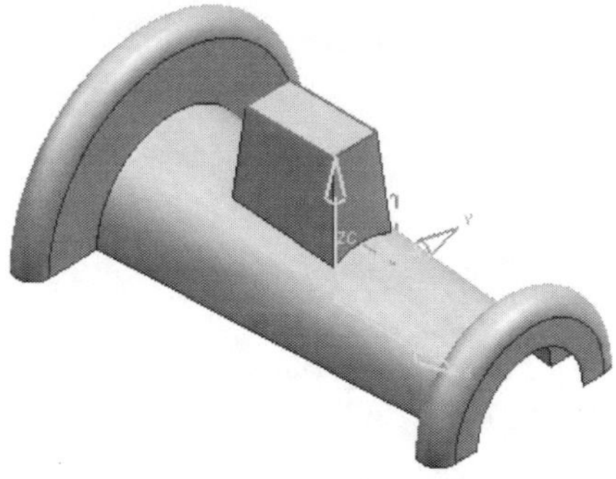

Figure 10-27 The resulting surface

Stitching the Extruded Surface and its Top Surface with the Revolved Surface

After creating all the surfaces, you need to stitch them together by using the **Sew** tool.

1. Choose the **Sew** tool from the **Menu > Insert > Combine** of the **Top Border Bar**; the **Sew** dialog box is displayed and you are prompted to select the target sheet to sew.

2. Select the revolved surface; you are prompted to select the tool sheets to sew.

3. Select the extruded surface and its top surface.

4. Choose the **OK** button; all the surfaces are stitched together.

Note

After stitching individual surfaces into a single surface, you can hide the sketch created for revolving the base feature, the sketch created for extruding the second feature, and the datum plane. To hide these entities, press CTRL+B; the ***Class Selection*** *dialog box is displayed. Select the entities and choose the* ***OK*** *button.*

Creating Fillets on Edges Using the Edge Blend Tool

1. Choose the **Edge Blend** tool from **Menu > Insert > Detail Feature** of the **Top Border Bar**; the **Edge Blend** dialog box is displayed and you are prompted to select the edges for a new set.

2. Select the edges of the surface, refer to Figure 10-28. Enter **5** as the fillet radius value in the **Radius 1** edit box.

3. Choose the **OK** button from the **Edge Blend** dialog box. The completed surface model after adding fillets to edges is shown in Figure 10-29.

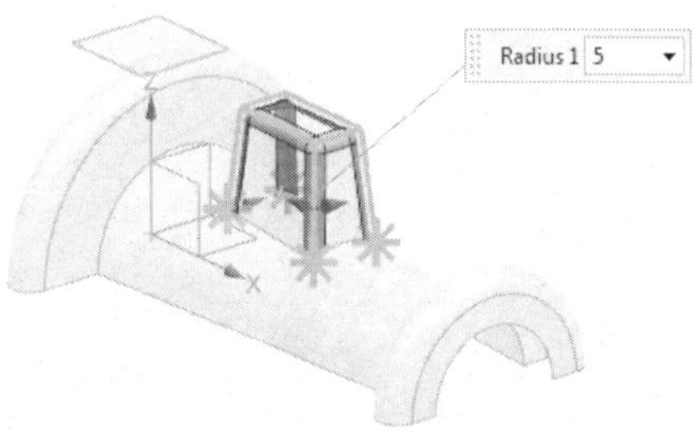

Figure 10-28 *The preview of the fillets displayed after selecting the edges*

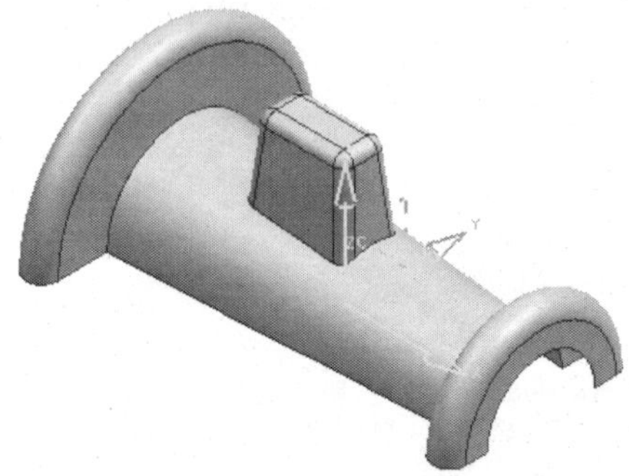

Figure 10-29 *The final surface model*

Saving and Closing the File

1. Choose **Menu > File > Close > Save and Close** from the **Top Border Bar** to save and close the file.

EXERCISES

Exercise 1

Create the surface model shown in Figure 10-30. The drawing views and dimensions of the surface model are shown in Figure 10-31. Save the model with the name *c10exr1.prt* at the location *\NX\c10.*

(Expected time: 30 min)

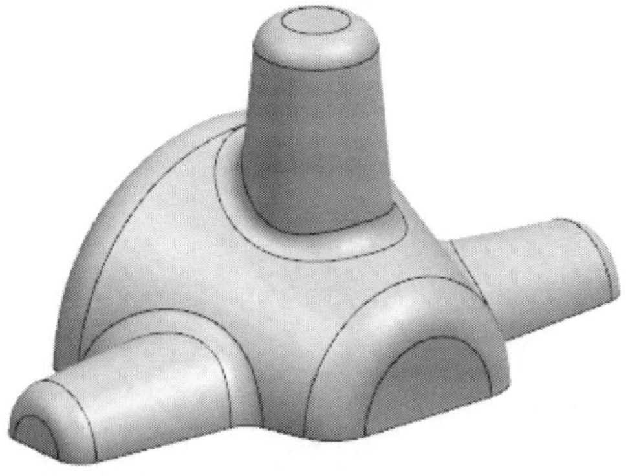

Figure 10-30 Surface model for Exercise 1

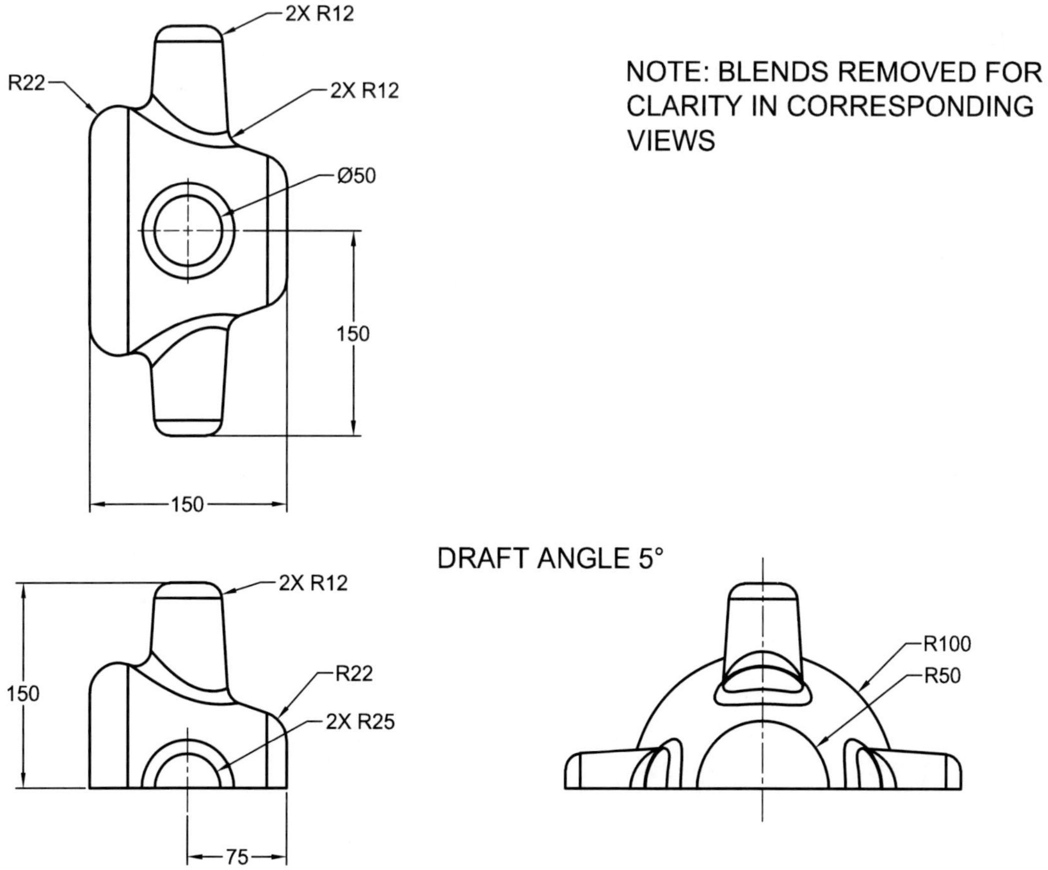

Figure 10-31 Drawing views and dimensions of the surface model for Exercise 1

Exercise 2

Create the surface model shown in Figure 10-32. The drawing views and dimensions of the surface model are shown in Figure 10-33. Save the model with the name *c10exr2.prt* at the location *\NX\c10.*

(Expected time: 30 min)

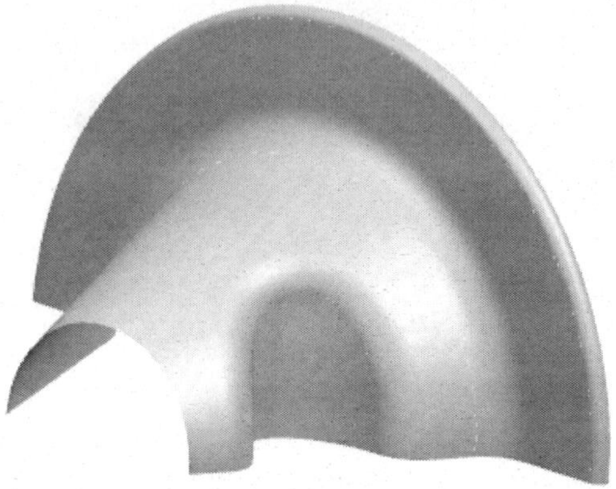

Figure 10-32 *The isometric view of the surface model for Exercise 2*

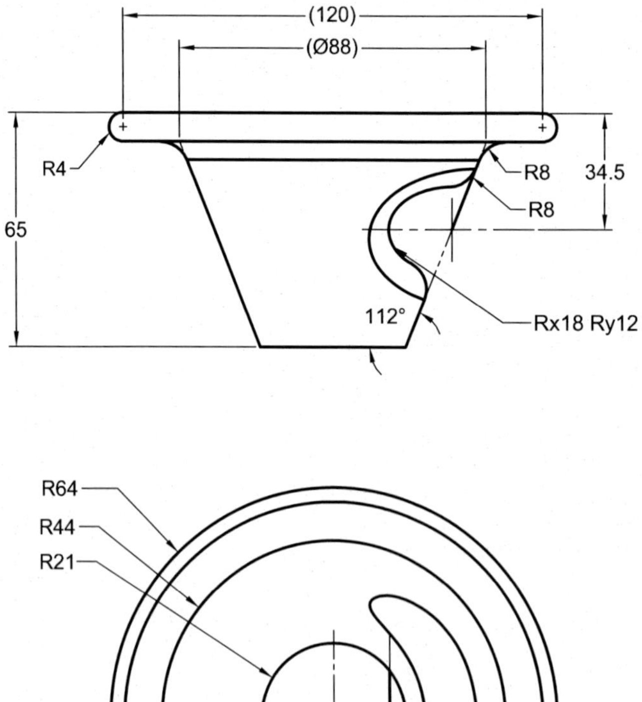

Figure 10-33 *The dimensions and drawing views of the surface model*

Index

A

Add Component 8-4
Arc tool 2-4
Bottom-up Assembly 8-3
Assembly Environment 1-3

B

Edge Blend 6-4

C

Chamfer dialog box 6-3, 6-4
Chamfer tool 6-3
Circle tool 2-4
Create New Component 9-2
Create New Component dialog box 9-3
Creating Features In the Assembly File 9-2
Datum CSYS 5-4

D

Datum CSYS dialog box 5-4
Draft dialog box 7-11
Drafting Environment 1-3

E

Edge Blend dialog box 6-4
Shape Studio Environment 10-2
Extrude 4-4

F

Feature-based Modeling 1-6

G

Getting Started With NX 1-4
Groove 6-3
Groove dialog box 6-3

H

Hole tool 6-2

I

Inferred tool 5-5
Introduction to NX 1-2
Sketching Tools 2-3

L

Line tool 2-3

M

Modeling Environment 1-2
Modeling Preferences dialog box 10-2
Surface Modeling 10-2

P

Parametric Modeling 1-6
Fixed (Principle) Datum Planes 5-3
Relative Datum Planes 5-3

Q

Quick Trim tool 4-2, 4-3

R

Rectangle 2-4
Reference Set 8-6
Revolve tool 4-4

S

Shape Studio Environment 1-3
Shell dialog box 7-10
Shell tool 7-9
Constraining Sketches 3-2
Dimensioning Sketches 3-3
Finish Sketch 2-4
Starting NX 2-2
Extruded Surface 10-3
Revolved Surface 10-3
Sweep along Guide dialog box 7-6

Sweep along Guide 7-5
System Requirements 1-4

T

Thicken dialog box 10-8
Thicken tool 10-8
Thread dialog box 7-8
Through Curves dialog box 10-7
Profile Tool 2-3
Top-down Assembly Design Approach 9-2

U

Using the Fillet tool 2-1

Made in United States
North Haven, CT
04 November 2026

21728191R00102